The End of
Darwinism

Also by Eugene G. Windchy

Tonkin Gulf

The End of Darwinism

And How a Flawed and Disastrous Theory Was Stolen and Sold

Eugene G. Windchy

Library of Congress Control Number:		2008910238
ISBN:	Hardcover	978-1-4363-8369-1
	Softcover	978-1-4363-8368-4

This book was printed in the United States of America.

To order additional copies of this book, contact:
Xlibris Corporation
1-888-795-4274
www.Xlibris.com
Orders@Xlibris.com
51516

CONTENTS

To all the Elizabeths

Acknowledgments

I am grateful to many prominent scientists for their assistance. They include David M. Raup, University of Chicago paleontologist; Steven M. Stanley, Johns Hopkins University paleontologist; John J. Sepkoski, University of Chicago paleontologist; Richard Potts, Smithsonian Institution paleoanthropologist; Thomas Gold, Cornell University astrophysicist; David Langebartel, University of Wisconsin anatomist; David Deck, University of Virginia embryologist; Kenneth J. Hsu, Swiss Federal Institute of Technology geologist; Lynn Margulis, Amherst University cell biologist; William Howells, Harvard University physical anthropologist; Marcel Schutzenberger, University of Paris mathematician and information theorist; Murray Eden, Massachusetts Institute of Technology mathematician and electrical engineer; David Jablonski, University of Chicago paleontologist. I regret that my research has taken so long that some of these helpful persons have passed away.

Special thanks go to a star librarian, Miss Constance Carter, at the Library of Congress. Miss Carter has a well earned reputation for being able to locate anything whatsoever within her vast realm. If somebody puts a book about the titmouse on the mammal shelf, she will find it.

Preface

"Let's Get Rid of Darwinism" urged a *New York Times* article of July 15, 2008. The author was a British evolutionary biologist, Olivia Judson, a regular contributor to the *Times*. Judson declared that she wanted the word *Darwinism* to be abolished because it—and the criticisms of it—conveyed too narrow a view of evolution. The British reformer said she would prefer to discuss more recent developments in evolutionary biology. It is not surprising that a biologist would tire of defending a nineteenth century theory that still is under attack. And for this biologist the task must be rather difficult because Judson herself entertains some doubt about the theory. As we shall see, the *Times's* evolution expert believes it possible that evolution has occurred not only in the way Charles Darwin theorized, but also in a different way which, according to Darwin, would have been miraculous.

In this book, we will discuss some recent and startling developments in evolutionary biology, but principally we must investigate basic Darwinism and its modern variant neo-Darwinism. They still inform the textbooks and much of our modern thinking.

The outspokenness of Olivia Judson might presage an era of change. When she ascended to her bully pulpit at the *Times*, two dominant figures in evolutionary biology had passed away. They were Ernst Mayr and Stephen J. Gould, both of Harvard University. I learned much from these two men, and I wish they were still alive. For years, I looked forward to seeing their reaction to my slowly prepared book. However, progress in science often depends upon the departure of the old guard, and evolutionary biology now could be on the brink of fresh and unfettered thinking.

Prologue

There's a strong school of thought in biology that one should never question Darwin in public. Computer scientist W. Daniel Hillis quoted in *The Third Culture* (edited by John Brockman), 1995

He who does not want to fight in this world of perpetual struggle does not deserve to live! Adolf Hitler quoted in the evolution section of German biology books, 1942

In 1852, seven years before *The Origin of Species* was published, a young zoologist, Thomas H. Huxley, was in London trying to make a career for himself and he wrote home to his sister Eliza: "You have no notion of the intrigues that go on in this blessed world of science. Science is, I fear, no purer than any other region of human activity; though it should be."

That observation, as I shall demonstrate abundantly, applies especially to the field of evolutionary biology. When Charles R. Darwin died, he was lauded for his "modesty," "honesty," and "simplicity." He even was extolled as a "saint in disguise." That was a carefully crafted image, burnished by influential friends, which survives to the present day. In reality, Darwin was a master of tact and charm, but underneath those polished manners lurked an intensely ambitious scientist who advanced his career by means of deception and intrigue. In that way he also advanced the theory which is attributed, incorrectly, to him.

For the sake of convenience, I shall use the commonly accepted terms *Darwin's theory* and *Darwinism* although Darwin was not the first to publish on evolution by natural selection. Two men preceded him. A third would have done so if he, working in a tropical jungle, had not sent his manuscript to Darwin for forwarding.

Two friends of Darwin, Ernst Haeckel and the aforementioned Thomas Huxley, became so famous as evolution advocates that mountains were named for them in California. However, neither Huxley, "Darwin's bulldog," nor Haeckel, "the German Darwin," actually believed in Darwin's theory, that is, his explanation of how evolution works. According to Darwin, evolution occurs very gradually by a combination of chance variation and natural selection (survival of the fittest), and old species become extinct as new and better adapted species out-compete them. Huxley doubted both gradualism and natural selection. Haeckel disbelieved in the role of chance.

Darwin's two major allies did agree that *some* kind of evolution took place—and with no Creator involved. Both men had complaints about the Christian church.

Haeckel was the leading exponent of evolution on the continent of Europe, and he was so enthusiastic about it that he forged evidence for human evolution. The German Darwin was convicted of fraud by a faculty court at the University of Jena, his employer in Prussia, but this did not prevent Haeckel, the author of best selling books, from continuing to spread misinformation. Neither did it prevent him from becoming an international apostle of materialism and a major source of Nazi philosophy.

Nazi Germany took very seriously survival of the fittest theory. German biology books drove the lesson home with a quotation from the nation's most prominent Darwinist, Adolf Hitler, who declared that he who does not want to fight, does not deserve to live.

Another eager recruit to Darwin's cause was his contemporary Karl Marx, a German scholar living in London, who received his doctorate from Jena. Marx loved Darwinism. To him survival of the fittest as the source of progress justified violence in bringing about social and political change, in other words, the revolution. Darwin "suits my purpose," Marx wrote to his comrade Ferdinand Lassalle.

Already we can see that Charles Darwin's theory, applied to politics, has helped to bring about an incalculable amount of suffering and death. Yet in much of the world's scientific community, including the United States, it for long has been unacceptable to criticize Darwinism in public, and at times such criticism has been deemed unacceptable even in the scientific literature. This militant, unscientific attitude came about as part of the Darwinists' fight against the creationists. Scientists, even those who are not biologists, tend to believe that it is important to maintain a united front in support of Darwinism.

The use of fraudulent materials did not end in the nineteenth century. It continued in the twentieth century and on into the twenty first. Amazingly, Haeckel's nineteenth century misrepresentations still can be seen in some American textbooks and reference books, even a book co-authored by a former president of the National Academy of Sciences, an organization that publishes guidelines for the teaching of evolution. The many mistakes and deceptions associated with Darwinism possess a Houdini-like ability to escape the self-correcting mechanisms of science. The continuance of Haeckel's forgeries a British scientist has blamed on the *historians*, saying that they ought to report such things to the scientists. I hope this book will prove useful along that line.

Errors and deliberate deceptions have done a great deal to promote acceptance of Darwinian theory, and so we must ask whether the theory is so well founded as we are told. The science establishment, the textbook writers, the education authorities, and our staunchly Darwinian judiciary—they all want Darwinism to dominate the teaching of biology, free from criticism. The ruling powers do not seem to communicate well with the scientists who do research in evolutionary biology. Many of the latter are critical of textbook Darwinism and some even have compared it to a religion.

One of the critics is David M. Raup, who at this time probably is the world's leading expert on fossils. Now retired from the University of Chicago, Raup is a former president of the Paleontological Society, a recipient of the Paleontological Society Medal, and a recipient of the National Science Medal. The famous evolutionary biologist Stephen J. Gould rated him "the best of the best." Raup agrees that natural selection does happen, but he questions the *importance* of natural selection—"whether it has been responsible for 90 percent of the change we see, or 9 percent, or .9 percent." That criticism goes to the very heart of Darwinian theory.

What could be wrong with the theory of natural selection? Why might not the struggle for existence shape evolutionary change? As one example, consider the dinosaur, a charismatic beast that fascinates our young people. New discoveries are challenging the old Darwinian explanations of how the dinosaur came about and why it became extinct. Fossils found in 2007 contradict, or at least do not support, the assumption that the dinosaur was shaped by a struggle for existence with a parental species. Neither was the king of the Jurassic rendered extinct because of a Darwinian failure to compete. It now is well established that the extinction of the dinosaur and most of its contemporary animals resulted from a cataclysmic event such as the impact of a comet or meteorite.

A more severe critic of Darwinism than Raup is the cell biologist Lynn Margulis, the world's leading authority on microbial evolution and a recipient of the National Science Medal. Margulis, noting a lack of evidence, compares the modern neo-Darwinism to a "religious sect." A similar comparison was made by Colin Patterson, senior paleontologist at the British Museum of Natural History. One of the most distinguished European evolutionary biologists of recent times was the Sorbonne's Pierre-Paul Grasse, the author of more than three hundred works and a president of France's National Academy of Sciences. Grasse labeled Darwinism a "pseudo-science."

Despite such criticisms, American textbooks on biology form a solid phalanx of survival of the fittest Darwinism, and concerned parents, if critical, are called "anti-science," "ignorant," or "religious fanatics." The British geneticist Richard Dawkins once got so riled up about people who say they disbelieve in evolution that in the *New York Times* of April 9, 1989, he branded them "stupid," "insane," and even "wicked."

In that atmosphere, how many historians would want to step forward and offer a correction?

It is not just Darwin advocates like Haeckel who have been fooling people. Darwin himself was good at that. Let us examine one of his most audacious deceptions. This nineteenth century ruse still goes on, and some years ago it tripped up even a prominent Darwinist, Carl Sagan, who was a frankly anti-religious astronomer and science popularizer. Unfortunately for Sagan this happened as he was debating on television with the Reverend Pat Robertson. I tuned in just as the astronomer was admonishing Robertson, "You have not met Mr. Charles Darwin."

The minister shot back, "Yes, I have met Mr. Charles Darwin, and he believed in God!"

Apparently unable to think of anything to say, Sagan just sat there and looked down at the floor. It always is dangerous for a scientist to pontificate outside his field, and here was a world famous one getting his head handed to him by a televangelist. Many a Christian minister might have done the same. If Sagan had been well informed on his subject, he could have stunned Robertson and probably millions of other Christians with the following sort of response:

"I am sorry to disappoint you, Reverend Robertson, but in fact Charles Darwin did not believe in God. He only pretended to. In a letter to a close friend he regretted having 'truckled to public opinion' by saying in *The Origin of Species* that the Creator was responsible for the first living organism. In his autobiography, which was published posthumously, Darwin revealed that he was agnostic."

Giving the Creator a role in *The Origin of Species* not only improved Darwin's reputation but made somewhat less objectionable his anti-Biblical theory. Most important, the Creator ruse helped the author to avoid questions about his most difficult problem, the origin of life.

Darwin never could find a scientific explanation for how life began, and to this day it remains a mystery. One of the most prominent researchers in that field has been the British molecular biologist Francis Crick. After Crick won a Nobel Prize for his co-discovery of DNA's double helix structure, he tried to figure out how life originated, and becoming very frustrated by that task, he complained, "Every time I write a paper on the origins of life I swear

I will never write another one, because there is too much speculation running after too few facts."

To report on Darwin's Creator trick more fully, the first edition of *The Origin of Species* came out in 1859, and concerning the origin of life it said vaguely, "There is grandeur in this view of life, with its several powers, having been originally breathed into a few forms or into one." That satisfied nobody, scientist or non-scientist. But soon there came out a second edition, and it said, "There is grandeur in this view of life, with its several powers, having been originally breathed by the *Creator* [emphasis added] into a few forms or into one."

That was not the author's true belief. In 1863 Darwin wrote to the botanist Joseph Hooker: "But I have long regretted that I truckled to public opinion, and used the Pentateuchal term of creation [actually "the Creator"], by which I really meant 'appeared' by some wholly unknown process." Darwin's autobiography came out five years after his death in 1882, but the admission of agnosticism was censored by the Darwin family for several decades. A granddaughter, Nora Barlow, at last published in 1958 what is said to be the full text. In there Darwin says, "The mystery of the beginning of all things is insoluble to us, and I for one must be content to remain an Agnostic."

The autobiography also says that at the time Darwin wrote *The Origin of Species* he believed in an intelligent "first cause." Darwin sometimes was not truthful about himself. But assuming the "first cause" statement was true, that belief evidently waned at least by 1863. Six editions of *The Origin of Species* were published, the last in 1872, but the author, despite his regrets and his agnosticism, ostensibly stuck to the Creator as the originator of life. Darwin found plenty of time for making other changes. The book grew in length by nearly one-third as the embattled evolutionist added information and responded to criticisms.

Of course, it would not have helped Carl Sagan's case for Darwinism to announce on national television that its founder was intellectually dishonest. Maybe that is why the astronomer looked at the floor. It is more likely, in my opinion, that he simply did not know that Darwin had sought divine help in jump-starting the evolutionary process. This fact is overlooked by our public institutions of education.

For example, in 2005 I visited the hugely popular Darwin exhibit at the American Museum of Natural History in New York City, and I noticed that when quoting Darwin on the origin of life, it left out the part about the Creator. The exhibit managed to do that by quoting from the first edition of *The Origin of Species* instead of the commonly available sixth edition. Customarily for scholarly purposes, the author's final text would be quoted. In this case quoting from the first edition actually gave a more accurate view of the author's thinking. It also avoided dealing with a deception that now has been going on for a hundred and fifty years.

The Darwinists, ever alert to defend themselves against the creationists, are very good at covering up or ignoring their leader's ethical lapses and other shortcomings. As the computer scientist W.

Daniel Hillis has observed, among biologists it is considered very bad form to question Darwin in public. Stephen Gould had commented on that subject long before. In *Discover* magazine of May, 1981, he reported a "trend" among his colleagues to "mute" debate on theory because it "provides grist for creationist mills, they say, even if only by distortion." Gould added, "Perhaps we should all lie low and rally round the flag of strict Darwinism, at least for the moment—a kind of old-time religion on our part." So as not to be accused of quoting Gould out of context, following is a paragraph that came right after the previous sentence and ended the article:

> But we should borrow another metaphor and recognize that we too have to tread a straight and narrow path, surrounded by roads to perdition. For if we ever begin to suppress our search to understand nature, to quench our own intellectual excitement, in a misguided effort to present a united front where it does not and should not exist, then we are truly lost.

Gould left his own position ambiguous, but clearly a cover-up was underway. Fourteen years later Hillis observed that the biologists' united front, their "old time religion," was still going strong.

When people talk about evolution, they usually mean the history of life as explained by Charles Darwin, and it is a common misconception that Darwin discovered evolution. He did not. The coming and going of life forms was known long before Charles Darwin was born. He was one of many scientists who have tried to explain evolution. Here I use the word *evolution* to mean change over

time (one of the dictionary definitions). In addition to discussing
Darwin's explanation, I shall take note of Lamarckism, panspermia,
mutationism, symbiogenesis, and intelligent design.

One of many evolutionary theorists was Charles Darwin's
grandfather, the medical doctor Erasmus Darwin. Erasmus published
his ideas in the 1790s, well before his grandson's birth in 1809 (sharing
a February 12 birthday with Abraham Lincoln). When in 1816
Mary Shelley wrote *Frankenstein*, she was inspired by rumors about
Dr. Erasmus Darwin, the notorious evolutionist, and his mysterious
laboratory, where he reputedly had created life. As one can see in the
Oxford English Dictionary, Erasmus Darwin's theory was the original
Darwinism. The term changed its meaning after his grandson Charles
in 1859 published *The Origin of Species.*

Disagreements and various theories to which the textbooks pay
little or no attention fill the specialized books written by evolutionary
biologists. Two such books are *One Long Argument* by Ernst Mayr
of Harvard University and *Reinventing Darwin* by Niles Eldredge of
the American Museum of Natural History. Mayr's book is a scholarly
history of conflicts on theory among evolutionary biologists in
general. Eldredge's book focuses on a rift between the geneticists,
who study genes, and the paleontologists, who study fossils.

A bigger rift exists between the Darwinists as a whole and their
critics in the field of mathematics. The latter, on the side of Haeckel,
have viewed as extremely unsound, if not preposterous, the Darwinian
reliance upon chance. We will look into a heated international

The university said that Langebartel had retired to Florida. Actually he had moved again, this time to Arizona, but with some effort I tracked him down. It turned out that Langebartel did not know about the false information in the biology textbooks. He mentioned the non-existence of gills so as to prevent misunderstanding that could result from that part of the embryo's being termed "branchial arch." *Branchial arch*, not *gill clefts* or *gill slits*, is the term used by the books on anatomy. The term derives from the Latin (and Greek) word for gill, and Langebartel feared that the derivation would give some students the wrong idea.

Langebartel, I'd say, deserves a prize for being an extraordinarily conscientious writer of textbooks.

I did not doubt the anatomist's word, but I thought before publishing this information I'd better check with an embryologist. One of those I reached at the University of Virginia, where Professor David Deck also said there were no gills on the human embryo. Informed that some biology books told of "gill slits," Deck responded flatly, "They are not slits." One such erroneous textbook is the college level *Biology* by Peter Raven and others. I refer here to the seventh edition (page 1095) which was published in 2005 by the firm of McGraw-Hill. Looking into *Gray's Anatomy* (Churchill and Livingstone's thirty-eighth edition) I found the term *branchial arch*.

Gould and others have complained about how textbook writers copy each other. Apparently when one makes a mistake, it spreads around like a computer virus, especially if it supports a fashionable theory.

Gould did find out about another bit of textbook nonsense: Darwin's explanation for the height of the giraffe. Beginning in 1988 the professor campaigned for removing the giraffe from textbooks, but despite Gould's considerable prestige it took many years for his efforts to have much effect, and Darwin's nineteenth century giraffe fantasy at this writing still appears in some textbooks.

Since most American adults living today have had this hoary nonsense foisted upon them, we shall discuss later the errors involved. They are elementary to the point of absurdity. For one thing, Darwin was misinformed about the feeding habits of the animal. Considering the longevity of this error, one would think that the giraffe was a mysterious denizen of dense jungle, not a loping giant of the open country that has been captured and studied since the time of the Egyptian pharaohs. The ancient Romans, too, kept giraffes. Today one usually can find them in the nearest zoo. Anybody can go and see what they eat.

Why has the giraffe been so little understood by our textbooks? Again, the self-correcting nature of science has fallen victim to the desire to prove a theory. We shall see a lot of that.

There has been fostered a general impression that when *The Origin of Species* came out, the book's facts and logic were so obviously correct that Darwinism was opposed only by theologians and their followers. This is far from true. The book was rejected immediately by virtually the entire scientific community. A quarter century later Thomas Huxley reminisced, "There is not the slightest doubt that, if a general council of the Church scientific had been held at that time, we should have been condemned by an overwhelming majority."

The anti-Darwinian criticism was such that of the 3,878 sentences in the first edition, 75 percent were rewritten from one to five times each by the time the sixth edition came out in 1872. Opposition came especially from those scientists who specialized in the fields of anatomy, geology, and paleontology. One critic was the Cambridge University professor Adam Sedgwick, who had taught the young Darwin about geology and paleontology. Sedgwick said Darwin's theory was unsupported by facts and made him laugh until his "sides were sore." In America Louis Aggasiz, the Ice Age authority, condemned the theory as "a scientific mistake, untrue in its facts, unscientific in its method, and mischievous in its tendency." England's "head of science," the astronomer Sir John Herschel, denounced Darwinism as the "law of higgledy-piggledy."

Disgustedly, Darwin wrote to his friend Hooker: "I see plainly that it will be a long uphill fight." Not content merely to publish his theory, Darwin wanted everybody to believe in it. Janet Browne, who wrote a two-volume biography of the scientist, commented, "Overnight, Darwin's focus shifted from private effort to public persuasion." This zeal for converts does not square with the long-established notion that Darwin delayed publishing for twenty years for fear of damaging the fabric of society.

As mentioned above, another of Darwin's critics happened to be his chief ally in Britain, Thomas Huxley. The famous zoologist who went out and sold evolution with magazine articles and lectures, the fierce debater, the man known as "Darwin's bulldog"—he did not really believe in Darwin's theory. Nonetheless, Huxley routinely is described as a champion of Darwinism, the theory's "first adherent," and Darwin's "chief supporter" (*Encyclopaedia Britannica*). "Darwin's bulldog" was a name Huxley gave to himself and it stuck. However, what Huxley consistently campaigned for was not Darwin's theory, which he soft-pedaled, but the principle of secular evolutionism as opposed to creationism.

Huxley's initial reaction to natural selection was, "How extremely stupid not to have thought of that!" But then he came to doubt the ability of natural selection to shape new species, although he felt new varieties, fertile with their parental species, were possible. Huxley noted a lack of evidence for the effectiveness of natural selection, and he doubted that a new variety could separate itself reproductively from its species.

A later skeptic, the Dutch botanist and mutation theorist Hugo de Vries, asserted that natural selection may explain the *survival* of the fittest but not the *arrival* of the fittest.

Darwin's notion of extremely gradual evolutionary change, similar to artificial breeding, Huxley did not accept, either, and he said so right away, advising Darwin, "you have loaded yourself with an unnecessary difficulty." Huxley could not see gradualism demonstrated by the fossils. The jumps from one species to the next were too big. They were not Darwin's "numerous, successive, slight modifications" or "insensibly fine steps" (but not so fine that an expert breeder could not see them). Darwin viewed evolution by jumps as akin to miracle. Transmutation, said he, had to be shaped very gradually by the effects of environment. In *The Origin of Species* Darwin acknowledged the gaps in the fossil record, but he blamed them on insufficient discovery. In other words, he discounted the available evidence in paleontology because it did not support his theory.

When lecturing to the public, Huxley dazzled the audience with science wonders, discussed the comings and goings of animal species, and sometimes explained the idea of natural selection. Faced with scientists, he watered down Darwin's theory to make it more acceptable. Darwin often was in poor health, and he preferred to stay at his country estate and let Huxley do the talking. The latter became known as the "pope" of evolution, the reclusive Darwin's vicar on earth, as it were, and, appropriately, he did as much interpreting as advocating. Darwin was annoyed by his surrogate's inconsistencies. After hearing one of Huxley's lectures, the recluse complained, "He gave no just idea of natural selection."

Huxley did not believe in Darwin's theory and he had no real theory of his own. Nevertheless, in his view of evolution there was no place for a Creator.

We find a parallel life in Stephen J. Gould. In the mass media he gave the impression of fighting for Darwinism, but, like Huxley, he wasn't really convinced, and in a scientific journal, *Paleobiology*

(Winter edition, 1980), he actually declared textbook Darwinism "dead." As a regular writer for *Natural History* magazine, Gould defended Darwinism in some ways and criticized it in other ways. These ambivalent articles stirred up the creationists, and that angered many of Gould's colleagues. But he was the mass media's go-to man on evolution, and Gould's support for Darwinism in that venue enabled him to escape a public excommunication.

As a case in point, in 1999 I was amused to see how Gould rode to the rescue of Darwinism in Kansas, where the state school board voted to let the local school boards decide what to teach about evolution. Apparently outraged by this attack on the Darwinists' united front, Gould wrote for *Time* magazine an essay titled, "Dorothy, It's Really Oz." In the text of the essay the Harvard professor skillfully ridiculed the Kansas plan, and he eloquently defended evolution as fact—while never mentioning Darwin or his theory. No doubt it was an editor who inserted the sketch of an angry-looking Darwin with a bump on his head. In effect, Gould defended textbook Darwinism, which he had declared dead, while not saying a word about it.

Sitting in Cambridge, Massachusetts, did Gould know what exactly he was defending in Kansas (maybe that pesky giraffe?) I could not find out. Wanting to obtain a list of the state's biology books, I telephoned long distance to the Kansas School Board, and a very uncooperative employee told me I would have to ask the individual schools—as if the board did not know what books it had authorized. The telephone company helped me to reach a high school, where a panicky biology teacher demanded, "Why are you singling me out?" Nothing would persuade the man to let me know what biology books were being used in his school.

How did public education get to be secret education? Sir Walter Scott understood that sort of thing. In his familiar words: "Oh what a tangled web we weave, when first we practise to deceive."

Darwin told Hooker that his reference to the Creator was a public relations gesture, and this surely was true. In Victorian times it was disgraceful in the extreme to be an atheist. Darwin was a rich country squire who prized respectability, and the public was suspicious of him. Furthermore, his wife, Emma, was religious, and her husband's work greatly troubled her. In view of all that, it was advantageous to feign a belief in the Almighty. In 1869, ten years after publication of *The Origin of Species*, Huxley, Darwin's utility man, invented the word *agnostic*. That helped evolutionists everywhere to ward off charges of atheism.

As mentioned above, there was another important reason for Darwin to call upon divine assistance. When he got to the very last paragraph of *The Origin of Species*, the author still had not explained the origin of life, and he did not know what to say. His theory was about "descent with modification." (Darwin preferred that term to *evolution*, whose Latin roots suggest an *unfolding*.) But descent from what? The scientist had painted himself into a corner. As Darwin saw it, this was unfair; the biologist ought not to be required to explain the origin of life any more than, in his opinion, the physicist ought to be required to explain the origin of matter (concerning which the physicists since have made a little progress). But the question was inescapable. Descent from what? Much later, in an 1871 letter to Hooker, Darwin fantasized life's having begun in "some warm little pond" as a chemical accident. But he despaired, "oh! what a big if!"

It did not help that over in France Louis Pasteur was disproving the spontaneous generation of life. This concept had been around since the time of Aristotle, and it actually was supported by scientific experimentation. A seventeenth century Belgian tested the claim that cheese and rags mixed together would produce mice. He put the required ingredients into a wooden cabinet, closed the door, and some days later mice indeed were discovered therein. The experimenter must not have known that mice can flatten themselves and squeeze through a crack. (I assume the experimenter knew about gnawing.)

The invention of the microscope for long *bolstered* spontaneous generation. In drops of water scientists could see bacteria apparently coming from nowhere. In a classic example of wrong-headedness, scientists observed bodily wounds generating bacteria—still more proof of spontaneous generation. A rival of Pasteur's, Archimede-Pouchet, believed that he had demonstrated the spontaneous generation of microbes, but as Pasteur could see, he had not provided completely sterile conditions.

In France spontaneous generation became a national controversy. It was perceived that life arising from non-life made unnecessary a Creator, and so the evolutionists rooted for Pouchet while the anti-evolutionists rooted for Pasteur. The latter was accused of bias on account of his belief in a Creator. But after Pasteur had been submitting papers on the subject for years, the French Academy of Sciences asked him and Pouchet to appear before an appointed commission and perform their experiments. Pouchet kept objecting to the arrangements, and so by default the academy in 1864 found in favor of Pasteur. This was considered a great victory for the anti-evolutionists; it also helped conservative Catholic politicians.

Unfortunately, in those days Pasteur's victory seemed not very important to the practitioners of medical science. Doctors needed to treat wounds aseptically, and they needed to deal with contagious diseases. But they were slow to learn from Pasteur, whom they regarded as a "mere chemist," and this prejudice contributed to thousands of deaths from infection during the Franco-Prussian War of 1870. (Even today hospital sanitation is a major health issue. "Our Unsanitary Hospitals" was the title of a *Wall Street Journal* op-ed of November 29, 2007, which said that restaurants in New York and Los Angeles are held to more rigorous standards of sanitation than hospitals. Nationally more than 100,000 people are said to die each year from infections contracted in health care facilities while only 2,500 deaths result from food-borne illnesses.)

To this day nobody has seen life come from non-life despite many efforts in the laboratory to make that happen, and the biology books remain stuck with spontaneous generation as the beginning of evolution. But they don't call it spontaneous generation. They prefer the Greek word *abiogenesis*, which sounds better. The *World Book Encyclopedia* (2006 edition) is candid on this subject. It says, "Today, most scientists believe that spontaneous generation took place at least once—when certain chemicals came together to form the first simple living organism more than three billion years ago."

Harvard University Professor George Wald, who won a Nobel Prize in physiology, defended spontaneous generation as the "reasonable view" since the "only alternative" was to "believe in a single, primary act of supernatural creation." Wald's defense of spontaneous generation appeared as an article, "The Origin of Life," in the August 1954 *Scientific American*. Yet in the next decade the magazine ridiculed the Creation Science movement by comparing its credibility to that of spontaneous generation.

The alleged phenomenon of abiogenesis often has been criticized on mathematical grounds. One such critic was the French biophysicist Lecomte du Nuoy, who in the 1940s calculated that there had not been enough time in the entire history of our planet for chemicals to have come together randomly and form a living organism, much less in just the early days. Once I heard Sir Julian Huxley, the grandson of Thomas Huxley, asked about Lecomte du Nuoy's assertion. Huxley threw up his hands, literally, and snorted that the Frenchman's work was "utterly unscientific" and not worthy of consideration. Huxley's audience of American college students groaned in embarrassment because one of their number had asked such a dumb question.

Dumb question or not, the academic situation is comical. The history books tell us that Pasteur disproved spontaneous generation while the biology books tell us that abiogenesis initiated the climb from microbe to man. The historians are going by Pasteur's evidence, and the biologists, uncharacteristically, are going by faith. Of course, it is not possible to prove that abiogenesis never has happened. That provides some wiggle room for the biologists, but believing in the equivalent of spontaneous generation ought not to be required in order to get a good grade on a test.

Darwinism, especially the natural selection part of it, has had a destructive effect on traditional ethics and values. After *The Origin of Species* was published, thoughtful people worried about the social and philosophical implications. Darwin's friend Sir Charles Lyell, the world's most prominent geologist, feared that if men were *told* they were beasts, they would *act* like beasts (as we see in much of today's popular music). Some citizen wrote to a newspaper in Manchester saying Darwin has proved "'might is right,' and therefore Napoleon [III] is right, and every cheating tradesman is also right." Darwin called the letter "rather a good squib." He seems to have taken the matter lightly. Yet as an opponent of slavery Darwin found "revolting" the might is right philosophy of a pro-slavery acquaintance, the historian Thomas Carlyle.

The man in Manchester raised a legitimate and grave concern. Darwinism valued only survival and self-interest, not honesty, generosity, loyalty, or compassion. The dramatist George Bernard

Shaw hated the theory for its "hideous fatalism." At the opposite pole from Shaw stands a twenty-first century professor of philosophy, Daniel C. Dennett, of Tufts University. Dennett esteems natural selection "the single best idea that anyone has ever had," and he characterizes the concept as a real life "universal acid," the hypothetical acid so powerful that no container can hold it. "Darwinism," exults the professor, "eats through just about every traditional concept and leaves in its wake a revolutionized world view."

And soon there were consequences. As George Bernard Shaw observed succinctly, Charles Darwin "had the luck to please everybody who had an axe to grind."

In order to whip up public support for the Franco-Prussian War, the militarists claimed that war improved the human species just as conflict in the wild improved the animals. Conversely, they said, the lack of warfare made a nation weak and effete. To be sure, there were other causes for the conflict, but it can be said that the Franco-Prussian War was the first to be blamed on Darwinism.

Meanwhile, rich industrialists embraced Social Darwinism, Herbert Spencer's philosophy which condemned trade union and government efforts to mitigate the ill effects of *laissez-faire* capitalism. Such measures impeded progress, said Spencer, by preserving the least fit. (It actually was Spencer who coined the term "survival of the fittest.") Social Darwinism was especially well received by the robber barons in the United States. When Spencer traveled to this country, one of his admirers, the steel tycoon Andrew Carnegie, persuaded him to visit the industrial city of Pittsburgh, which was not a popular tourist destination.

The steel-making capital of America, where men worked twelve hours a day and seven days a week, was a fine laboratory for Social Darwinism. But Spencer did not have the stomach for it. Repelled by the smoke, heat, and unearthly din, the frail intellectual averred that six months in that foul bedlam would be enough to justify suicide. In order to get out of town as quickly as possible, the philosopher broke his rule of always staying in hotels and escorted by Carnegie, he fled to the comfortable home of the tycoon's brother.

In New York City the Englishman was feted royally by industrialists who declared themselves an aristocracy of merit chosen by natural selection. Spencer was appalled by how his hosts had taken his philosophy and run with it, and he urged them to slow down. Life was not made for work, admonished the founder of Social Darwinism, work was made for life.

Andrew Carnegie believed in Spencer, but he had enough traditional values left in him to regard his fortune as a moral responsibility, and he gave away most of it, founding libraries and supporting other worthy causes. The richest man in the world's philanthropy did not include handing out money to the poor. They in his opinion would simply waste it. As a good Darwinist Carnegie focused his efforts on improving the minds of the "English speaking race."

Darwinism encouraged racism. Here is the original, full title of Darwin's book: *On the Origin of Species by Means of Natural Selection, or, The Preservation of Favoured Races in the Struggle for Life.* Darwin believed that whole races of humanity inevitably would be wiped out by natural selection. Said he in a letter of July 3, 1881:

> The more civilised, so-called Caucasian races have beaten the Turkish hollow in the struggle for existence. Looking to the world at no very distant date, what an endless number of the lower races will have been eliminated by the higher civilised races throughout the world.

Darwin thought the Maori people of New Zealand already were on the way to extinction, but that turned out to be incorrect, and nowadays the Turks too are thriving. Matters turned out very differently in Tierra del Fuego, another group of islands that Darwin visited during a voyage aboard the HMS *Beagle* as the ship's naturalist. The indigenous Fuegians did die out, the victims of European diseases. Perhaps their fate would have been different if the islanders had not massacred their missionaries, who otherwise would have provided medical help.

Darwinism got the Nazis into eugenics, and that led to the deliberate murder of six million people. Groundwork for the Nazi racial philosophy was laid not only by Darwin but by his cousin, Francis Galton, who coined the word *eugenics*; by Ernst Haeckel, the German Darwin; and by the Darwin-influenced philosopher Friedrich Nietzsche, who declared that God was dead. Both Haeckel and Nietzsche were exponents of militarism. Haeckel we shall discuss later in part because of his phenomenal success as a forger of scientific illustrations, which included gills on the human embryo.

Although born to a Catholic family, Hitler became a hard-eyed Darwinist who saw life as a constant struggle between the strong and the weak. Hitler's Darwinism was so extreme that he thought it would have been better for the world if the Muslims had won the eighth century Battle of Tours, which stopped the Arab advance into France. Had the Christians lost, he reasoned, the Germanic peoples would have acquired a more warlike creed and, because of their natural superiority, would have become the leaders of an Islamic empire.

Karl Marx commented to his friend Friedrich Engels on how Darwin recognized among beasts and plants his English society with its "division of labor, competition, opening up of new markets, inventions, and the Malthusian struggle for existence." Marx perceived Darwinism as a reflection of capitalism, and as noted above, thanks to Darwinism the bearded revolutionary could claim that his theory of inevitable class conflict had a basis in natural history. A journalist compared Marx's work to Darwin's, and the revolutionary responded,

> Nothing gives me greater pleasure than to have my name thus linked onto Darwin's. His wonderful work makes my own absolutely impregnable. Darwin may not know it, but he belongs to the Social Revolution.

Marx's followers in Russia, China, and elsewhere would far surpass the Nazis when it came to deliberate murder. The Communists have massacred scores of millions in the name of progress and their allegedly scientific ideology.

When Darwin died in 1882, a generation after *The Origin of Species* was published, the people of Great Britain, despite the bitterness of the evolution controversy, were taking pride in having produced a world-shaking scientist. It enhanced their conception of the British Empire as a spreader of civilization and enlightenment.

The South American Missionary Society claimed Darwin as an honorary member, and that was taken to show that the man's heart and faith were in the right place. The evolutionist indeed had accepted an honorary membership. The invitation came about because of his financial contributions; Darwin approved of how the missionaries discouraged human sacrifice, cannibalism, and licentiousness. The honorary membership made it possible to laud the decedent, who had acknowledged the Creator in his most famous book, as a "true Christian gentleman" and even a "secular saint."

Unhappily, Darwin had proposed the ruthless mechanism of natural selection as the designer of living things. But as an Anglican canon explained, that principle, "rightly understood," was not really inconsistent with Christianity because it "acted under the Divine Intelligence." Such theologizing and a general outpouring of adulation helped to facilitate a great coup by Darwin's supporters: the interment of their leader at Westminster Abbey. Not only that, but the remains were placed near that great man of science Sir Isaac

Chapter One

Charles Darwin Steals a Theory

Darwin engaged in what Leonard Huxley called "a delicate arrangement," the greatest conspiracy in the annals of science. Arnold Brackman, *A Delicate Arrangement,* 1980

The book called The Origin of Species *is not really on that subject.* Old joke among paleontologists

From its beginning evolutionism has had its dark side. An early evolutionist was the Greek philosopher Empedocles, who lived in the fifth century B.C. According to legend, Empedocles, conscious of death approaching, took his own life by jumping into the crater of Mount Etna. He did so in secret hoping that a sudden disappearance would persuade his followers that he had become a god. The trick was discovered when the volcano threw up one of the philosopher's distinctive bronze sandals.

On a more positive note, I can report that Empedocles refused a kingship, that he believed in democracy, and that he was a vegetarian who claimed that eating meat brought about war, murder, and cannibalism.

As for his evolutionary theory, Empedocles conceived that the first living beings were independent organs that came together by the attraction of love to form animals and human beings. Some of these creatures were monstrous and unfit for survival, and they disappeared. There we see an early version of natural selection.

An interesting form of secularism cropped up in the first century B.C. The Roman poet-philosopher Lucretius decided that the gods did not create the universe, that they lived off by themselves, and that they took no interest in human affairs. Existence, said Lucretius, is without beginning or end, there is no spirit, and everything consists of atoms. As evidence for the existence of atoms, he pointed out that rocks invisibly wear smooth in a stream. The poet theorized that atoms move through an infinite void, and that a swerve in their progress brought about individual objects such as stars and human beings. (In that way Lucretius anticipated the mysterious "ripples" in the radiation from the Big Bang that made possible the formation of the stars and everything else.) Even the gods were made of atoms, said the poet, but those atoms were smaller and finer than ours.

A love potion drove Lucretius insane, and he killed himself around 50 B.C.

Skipping over a number of other thinkers, we come to Dr. Erasmus Darwin, the colorful grandfather of Charles Darwin. Although he faked his M.D. degree, Erasmus did study medicine for more than two years, and he was so successful as a doctor that King George III tried to recruit him as his personal physician. But Dr. Darwin did not want to live in London, and besides, he was very busy. In addition to practicing medicine, he was a poet, a scientist, an inventor, an educator, a feminist with an eye for the ladies, and an all-around intellectual. Benjamin Franklin lived in England at this time and the two, having much in common, became friends.

Like Lucretius, Erasmus Darwin published his theory in Latin as a poem. According to the doctor, all life began with a single filament; evolutionary change occurs continuously; use or disuse of an organ can alter it; and mere *striving* can bring about change. Dr. Darwin's concept of "warring nature" possibly hinted at natural selection, a concept which, if made explicit, would have been a sacrilege dangerous to publish in the eighteenth century. The *Dictionary of Scientific Biography* credits Erasmus Darwin with the "creation of the first consistent, all-embracing hypothesis of evolution."

In those days the Italian Luigi Galvani discovered that electricity would cause a twitch in the leg of a dead frog, and scientists began to wonder whether electricity might even "galvanize" to life a dead brain. That intrigued the eighteen-year-old Mary Shelley, and so when she heard that a piece of vermicelli had vivified in the laboratory of Dr. Darwin, she was inspired to create the fictional Dr. Frankenstein and the creature that he galvanized to life.

Long after Galvani scientists continued to look for life-giving properties in electricity. When Charles Darwin fantasized life originating in a warm pond, he thought electricity, presumably a lightning stroke, might have aided the event. In 1952, two American scientists, Stanley L. Miller and Harold C. Urey, tried such an experiment. They electrified a mixture of gases that they believed (erroneously) had been present in the atmosphere of the early Earth, and the result was some amino acids, the building blocks of protein. Although the experiment did not produce life, it was considered extremely important at the time, and many other scientists, including Carl Sagan, tried electrifying mixtures of chemicals. But still no life. Looking back at the 1952 experiment, Miller said, "It looked very, very easy then."

In the first decade of the nineteenth century, a French naturalist, Jean-Baptiste Lamarck, published a theory suspiciously similar to that of Erasmus Darwin. Lamarck too wrote about the effect of striving, use, and disuse. He believed that nature had a self-perfecting tendency. Lamarck said the giraffe became tall because of his striving to reach high on the tree for edible leaves. Predatory birds acquired keen eyesight as they searched for prey, and fish in caves went blind from not being able to use their eyes. Such physiological changes, said the Frenchman, could be inherited.

As evidence for the inheritance of acquired characteristics, it is a fact that the ostrich is born with callosities on his body exactly where he touches the ground when resting. Similarly, the African warthog is born with callosities on his forelegs exactly where he rests when feeding.

Modern biology textbooks ignore Erasmus Darwin, and they mention Lamarck only to condemn the theory which they attribute to him. Of course, the notion of striving as a cause of physical change sounds like magic. As for inheriting the effects of use and disuse, that also is flatly rejected by textbooks, even though Charles Darwin accepted such forms of evolution. Experiments have failed to demonstrate the inheritance of acquired characteristics.

Concerning Lamarck the geneticist C. H. Waddington said in his book *The Evolution of an Evolutionist*:

> Lamarck is the only major figure in the history of biology whose name has become, to all intents and purposes, a term of abuse. Most scientists' contributions are fated to be outgrown, but very few authors have written works which, two centuries later, are still rejected with an indignation so intense that the sceptic may suspect something akin to an uneasy conscience.

The man who invented the word *biology* fell afoul of Darwinian militancy, and he still is being traduced. I will show, however, that a present day scientist at Harvard University apparently gets away with using Lamarckism to explain the origin of the bird.

Some readers might find Darwin's acceptance of use and disuse as a cause of evolution hard to believe. However, when a contemporary critic said that his theory relied entirely upon natural selection, Darwin wrote an indignant letter to the *Nature* scientific journal saying, "No one has brought forward so many observations on the effects of use and disuse of parts, as I have done." The letter would have been far more indignant if Huxley had not urged restraint. The first edition of *The Origin of Species* put the most emphasis on natural selection.

But the author lost some confidence in that, and in later editions he backslid into what now is called Lamarckism.

Strangely, Darwin claimed that he did not find useful the work of either his grandfather or Lamarck. In a letter to his friend Sir Charles Lyell, Darwin said that Lamarck's work "appeared to me extremely poor; I got not a fact or idea from it." In his autobiography Darwin said he had read his grandfather's book *Zoonomia* but without [its] producing any effect on me." Yet when Charles started writing a his own book on evolution, he first titled it *Zoonomia*.

Another slighting of Erasmus occurred in *The Origin of Species*. In later editions the author reviewed contributions to evolutionary thought from more than a score of other persons without mentioning his paternal grandfather except in a footnote. There he said, "It is curious how largely my grandfather, Dr. Erasmus Darwin, anticipated the views and erroneous grounds of opinion of Lamarck in his 'Zoonomia'."

Lamarckism Darwin discussed in the main text, which, incorrectly, identified the Frenchman as "the first man whose calculations on the subject [of evolution] excited much attention." The novelist Samuel Butler had something to say about that. Upon reading *The Origin of Species* Butler had become a convinced Darwinist, but later he decided that natural selection was just a piece of meaningless humbug. Erasmus Darwin, he thought, had published the better theory. Besides disagreeing with Charles Darwin's theory, Butler in his nonfiction book *Evolution, Old and New* criticized the scientist himself for his failure to credit properly the work of his grandfather. This led to a series of events in which Butler accused Darwin of using an ostensibly impartial German writer to get back at him.

The novelist had evidence. In 1879 Darwin arranged the translation and publication of a German magazine article about his grandfather, and he made a derogatory comment about Butler to the article's author, Ernst Krause. Krause then revised the German text of the article so that the English version would reflect unfavorably upon Butler for having praised Erasmus's theory. The bulk of Krause's essay was a glowing tribute to Erasmus Darwin, but incongruously, he concluded by saying that attempts to revive the doctor's theory show "a weakness of thought and a mental anachronism which no one can envy."

Darwin contributed a preface that read as if the article had been prepared prior to Butler's attack on him. This was true of the original German text but not true of the English version, and Butler was able to prove that by the contents. Darwin explained to Butler that alterations in translation were "so common a practice that it never occurred to me to state that the article had been modified." That did not satisfy his critic, who wrote a letter of complaint to the *Athenaeum* magazine. Darwin wanted to respond that the printer inadvertently had omitted a statement that the text had been revised, but Huxley persuaded him just to keep quiet.

In a later nonfiction book, *Unconscious Memory*, Butler again raked Darwin over the coals. Among other complaints he said that his enemy's writing was full of ambiguity and loopholes. Although eloquent at times, Darwin, whether by chance or design, did have a muddled, ambiguous way of writing about his theory. The Darwinian ecologist Garrett Hardin described the man's prose as "by turn clear, obscure, explicit, cryptic, suggestive, they have within them all the characteristics that litterateurs seek in James Joyce." Huxley also was emphatic on that point, but to quote him would involve an ethnic slur. Darwin claimed he couldn't work harder on "clearness" if he were a slave under the lash. Emma helped with the editing and so did Mr. and Mrs. Hooker. One of Emma's friends, Georgina Toller, enhanced the author's literary style and Darwin rewarded her with a silver vase. In later years daughter Henrietta became the family editor.

Be all that as it may, Ernst Mayr in his book *Animal Species and Evolution* remarked, "It has been claimed, not without justification, that one can find support in Darwin's writings for almost any theory of evolution."

Darwin himself did not respond to Butler's new attack, but a friend, George Romanes, fired off a letter to *Nature.* The X Club had a close connection with *Nature,* and Romanes was allowed to publish an *ad hominem* attack, which disparaged Butler as a "nobody" and an "upstart ignoramus" seeking attention. The novelist had to threaten legal action in order to get a response published. Butler then retorted that as yet no one had answered his charges.

Although Darwin once complained in *Nature* about a misinterpretation of his theory, as a rule he let other people fight his battles, and that enabled him to project a benign image quite different from the selfish, deceptive, and spiteful one perceived by Samuel Butler. "There never was a more honest man," said the London *Times* when the scientist died. The paper also admired his "childlike simplicity."

Charles Robert Darwin became an evolutionist around 1837 but kept quiet about that because evolutionists were held in low regard. By his own accounts, he one or two years later conceived of evolution by natural selection. He still kept mum.

By 1838 two other men already had published on natural selection. The first was a Scottish landowner, Patrick Matthew, who wrote a book titled *On Naval Timber and Arboriculture*. The book came out in 1831 when Darwin was just graduating from college. Matthew said that a species of tree in a nursery varies more than it does in the wild, and he explained that in the wild trees varying from the norm were not likely to survive. A "natural process of selection," he decided, kept the species uniform. Thus in comparing well cared for nursery trees to forest trees, Matthew perceived natural selection as having a *conservative* effect. However, it occurred to this canny Scot that, with a change in circumstances, natural selection might bring about an evolutionary effect. Said Matthew: [The] "progeny of the same parents, under great differences of circumstance, might, in several generations, even become distinct species, incapable of co-reproduction."

Did Darwin know what Matthew had published? Darwin biographer Loren Eiseley found a similar comment about nursery trees and forest trees in a Darwin essay of 1844. This essay was not published until long after the author's death; son Francis brought it to light in 1909. Eiseley thought it likely that Darwin had read Matthew's book and that he had derived from it the term *natural selection*.

After *The Origin of Species* was published, Matthew in a periodical wrote again about his concept. Darwin denied having known about the man's earlier work, but he mentioned it in subsequent editions of his book. Ernst Mayr said Matthew "had the right idea." Unlike Darwin, however, Matthew viewed nature as essentially benevolent, and he believed that beauty came from the Creator.

The second man to write about natural selection was the English naturalist Edward Blyth. Blyth in 1835 and 1837 published articles in *The Annals and Magazine of Natural History* in which he discussed variation, the struggle for existence, and the possibility of those conditions bringing about new and better adapted organisms. In the 1835 paper Blyth decided that the struggle for existence would exert a conservative effect. But in the second paper Blyth raised the question, "May not, then, a large proportion of what are considered species have descended from a common parentage?" On balance, though, he favored the conservative effect.

Eiseley complained that Darwin credited Blyth for some parts of his work but never for the concept of natural selection.

Darwin's third rival started publishing on evolution in 1855. This precipitated a case of deception and politicking that, amazingly, still goes on. First, let us examine the background.

Famously, Darwin saw how bird species had diverged on the Galapagos Islands from their presumed ancestors in South America, and so it was no great intellectual leap for the grandson of Erasmus Darwin to become an evolutionist. But it was emotionally wrenching. The young scientist felt so guilty about his conversion from creationism that acknowledging it was "like confessing a murder," as he wrote to Joseph Hooker on January 11, 1844.

The guilt stemmed from his religious training; at Cambridge University the future evolutionist had studied for the Anglican priesthood. This he did because he was a believing Christian, because his father demanded that he have a career, and because as a country rector he would have plenty of spare time for tramping around in the outdoors and studying natural history. The young man had started out to be a physician like his father and grandfather, but at medical school he encountered the horrors of pre-anesthesia surgery and dropped out.

Charles must have felt guilty about his failure to become a doctor. Dr. Robert Darwin viewed his bird-shooting, beetle-collecting son as a wastrel, and this disapproval came with an extra-sharp sting because the older man was generally acknowledged to be a keen judge of character. Anecdotes were told about that. The doctor, for example, could sense who would meet a financial obligation and who would not. Charles called his father "the wisest man I ever knew" and remarked that his ability to read character and even thoughts was "almost supernatural." Physically Robert Darwin also happened to be one of the biggest men in England. The family joked that when he returned home, it was "like the tide coming in."

After Charles's graduation from Cambridge, a five-year voyage on the HMS *Beagle* as ship's naturalist carried the young theologian around the world, and the astute scientific observations he published carried him from a religious career into a scientific one.

How the trip came about is a study in irony. While planning a long voyage of hydrographic surveying, the captain of the *Beagle*, Robert Fitzroy, wanted to find a civilian dining companion in order to relieve the loneliness of command. Fitzroy's predecessor on the *Beagle* had committed suicide, Fitzroy's uncle had killed himself, and so the skipper might well have been concerned about such a possibility for himself (he eventually did die by his own hand). Helping Darwin to get the job was the notion, in the captain's mind, that a theologian who also was an ardent naturalist might be able on shore expeditions to find evidence for the creation story in Genesis, which was coming under attack from geologists and evolutionists. To help Darwin with one of the problems of creationism, the captain advised him that the mastodon became extinct because it was too big for the door of the Ark.

One thing about the candidate from Cambridge gave the skipper pause. As a student of physiognomy, the divining of character by the characteristics of the face, Fitzroy noticed that Darwin had a weak sort of nose. But in the end he went with common sense and brought the clever young Cambridge man aboard. So far as the Genesis project was concerned, the skipper would have done better to stick with his "bumpology."

Ship's naturalist was an unofficial position that required Darwin to pay his own expenses. That he did with ease, and he also hired an assistant, a teenaged fiddler by the name of Syms Covington, who later settled in Australia and from time to time continued collecting specimens for his *Beagle* employer. Darwin's father not only was a prosperous physician but had married an heiress from the Wedgwood family known worldwide for its manufacture of pottery. The Darwins and Wedgwoods were close, and they became even closer when, following the *Beagle* voyage, Charles married his cousin Emma, who also was a Wedgwood. (The tradition continued, and in the next generation a Wedgwood boy joked, "I never knew a Darwin who wasn't mostly a Wedgwood.")

Money never was a problem for Charles Darwin. That is how he was able to devote himself full time to scientific research even though in the nineteenth century science was not a paying profession. Many scientists were very poor. Edward Blyth barely survived on 250 rupees per month as curator for the Royal Asiatic Society of Bengal. Other scientists were men of some leisure. These tended to be independently wealthy or clergymen. The latter sought to learn about God by studying His creations. Among them were the monk Gregor Mendel, who discovered laws of heredity, and the chemist Joseph Priestley, a

Christian minister, who derived oxygen and carbon dioxide from the air. Priestley, the inventor of soda water, had been inspired to take up science by his acquaintance Benjamin Franklin.

Incidentally, a descendant of Franklin, William B. Scott, was a vertebrate paleontologist who spent sixty-seven years on the faculty of Princeton University. Scott died in 1947 without, according to him, having learned anything whatsoever about the causes of evolution.

Before publishing on evolution, Darwin became a prominent scientist in other areas of research. In 1839 he was elected to the Royal Society, and in 1853 the society awarded him the Royal Medal for his work on barnacles, coral reefs, and volcanism. (Among Darwin's descendants at least nine became fellows of the Royal Society.)

By approximately 1851 Darwin lost his Christian faith. This followed the deaths of his admired father and his favorite child, ten-year-old Anne Elizabeth. When the affectionate "little Annie" died (various diseases have been blamed), her father was so devastated that he went to bed and could not attend the funeral. For reasons of health Charles did not attend his father's funeral either. Some writers have attributed the man's sensitivity to the circumstance that his mother died when he was eight years old and the boy was raised largely by his older sister Caroline.

Even before those tragic events the cruelties of nature, which Darwin studied all the time, were eroding the scientist's religious faith. Most people find contact with nature soothing, therapeutic, or even "Good for the soul!" as a Scottish gardener once informed me with gusto. But Darwin did not focus on the beauties of nature; he examined in detail the ugliness underlying it. As he wrote to an American friend, the Harvard University botanist Asa Gray, the English naturalist could not believe that an omnipotent and benevolent God would want cats to play with mice or the larvae of wasps to feed on the living bodies of captive caterpillars. Censored from Darwin's autobiography when first published was the following sentence:

> A being so powerful & so full of knowledge as a God who could create the universe, is to our finite minds omnipotent & omniscient, & it revolts our understanding to suppose that his benevolence is not unbounded, for what advantage can there be in the sufferings of millions of the lower animals throughout almost endless time?

Darwin's reasons for disbelieving in a Creator were not scientific but emotional and theological. In a letter to Hooker he made this prophetic declaration: "What a book a Devil's Chaplain might write on the clumsy, wasteful, blundering, low & horridly cruel works of nature."

The more research he did, the more Darwin's health deteriorated. During his *Beagle* days the ship's naturalist could outmarch the sailors when ashore; at home he had hiked as much as thirty miles a day studying natural phenomena. Back in England something went wrong. He became melancholy, withdrawn, and sickly. He was very sensitive to changes in temperature and when indoors often wore a shawl (like Abraham Lincoln with whom he shared his birth date). The chief symptom of illness was stomach pain with frequent retching or vomiting. Over the years the patient also suffered from extreme flatulence, boils, dizziness, headaches, shivering, hysterical crying, dying sensations, eczema, faintness, and ringing of the ears. The scientist became more and more reclusive, shunning outside social engagements, which made him shiver violently.

Doctors have speculated on both psychological and physical causes for Darwin's ailments. Among the physical possibilities they have cited the parasitic Chagas' disease, which he might have contracted when exploring the wilds of Argentina. He in fact was bitten by the bloodsucking insect that can transmit the disease. The Chagas' disease explanation does not seem consistent with the remedy that Darwin found. He occasionally was able to improve his condition by getting away from his work to relax at a spa that offered the then fashionable hydrotherapy. Visiting by himself these luxurious establishments, the patient took nature walks, played billiards, made friends, and even enjoyed dining with the other guests. One of his favorite companions was the young romance novelist Georgiana Craik, who often visited the Moor Park spa when Darwin was there.

Although Darwin kept his evolutionary ideas secret from other scientists, excepting to some extent his friends Lyell and Hooker, he discussed them freely with Miss Craik as early as 1857. The novelist challenged the scientist on an admitted problem with his theory, gaps in the fossil record, but he liked her anyway. Coincidentally,

Darwin loved romantic fiction, a foible which the family considered embarrassing and pathetic. A favorite form of relaxation for this dour investigator was to lie back and have his wife read to him a book about the tribulations of some pretty heroine. There had to be a happy ending; otherwise the listener felt cheated and dissatisfied. This was Darwin's way of putting his mind at rest.

Another spa companion was Mary Butler, a charming teller of ghost stories. Darwin gave her autographs of famous naturalists, having torn the signatures from mail that he received. At least two spa visits were timed to coincide with those of this Irish lady. Following is an excerpt from a September 11, 1859, letter to Miss Butler written as he was finishing *The Origin of Species:*

> My book at last is so nearly finished that I can really & truly see that I shall be a free man at the end of this month. Our plans are rather undecided; but I incline strongly to go to Ilkley [a new and relatively distant spa], but I fear . . . that it is too late to take a house for my family; & in this case I should stop three or four weeks in the establishment, return home for a week or so, & then go to Moor Park for a few weeks, so as altogether to get a good dose of Hydrotherapy.
>
> My object in troubling you . . . is to know whether there is any chance of your being at Ilkley in the beginning of October. It would be rather terrible to go into the great place & not know a soul.

Although alleviated remarkably by visits to the spas, Darwin's illnesses were real or at least very troubling. When writing *The Origin of Species,* he was afflicted with vomiting and a severe case of eczema.

A speech to a scientific society once cost him "23 hours vomiting," as Darwin wrote to Hooker.

Most of the scientist's children became hypochondriacs in imitation of their father. That was the opinion of a granddaughter, the artist Gwen Raverat, who said that to be ill in the Darwin household was a "distinction" and a "mournful pleasure." Henrietta, who once had typhoid, was the most bedridden. The health of Raverat's father, George Darwin, improved after he married an American woman who did not enjoy nursing him.

The strong and pious Emma was the rock to which everybody clung. Yet she had her own cause for distress: the knowledge of her husband's theorizing and his apostasy. He kept "putting God further off." Many studies have shown that a strong religious faith is good for one's health and happiness. Did the evolutionist sense that? Once after a week of relaxing at a spa, Darwin rejoiced in a letter to Hooker, it is "quite unaccountable. I can walk & eat like a hearty Christian; & even my nights are good."

At the age of sixty-one Darwin encountered a hearty Christian while visiting his *alma mater*. This was the anti-Darwinian Professor Adam Sedgwick, who greeted his old student as joyfully as if *The Origin of Species* had never been published. Sedgwick insisted on giving his visitor a tour of the university's geological museum, and the experience exhausted the younger man, who later commented, "Is it not humiliating to be thus killed by a man of eighty-six, who evidently never dreamed that he was killing me?"

Following Edward Blyth the next scientist to publish something very significant on evolution was Alfred Russel (*sic*) Wallace, a young Welsh naturalist (of Scottish and English ancestry) who supported his research in the jungles of the Malay Archipelago by finding biological specimens of interest and selling them by mail order. Darwin was one of his customers. In 1855 the Welshman published in the *Annals and Magazine of Natural History* a major piece of evidence for evolution. The article was titled, "On the Law which has regulated the Introduction of New Species," and it stated the principle, "Every species has come into existence coincident in time and space with a closely allied species." This became known as the Sarawak Law since Wallace had written the paper while in Sarawak, Borneo.

Wallace listed ten facts in support of his law. Fact No. 8 said that those species of a genus occurring in the same geological time were more closely related than the species of the same genus occurring in a later geological time. In other words, species of a genus diverged in time. In the margin next to Fact No. 8 Darwin wrote, "Can this be true?" Evidently something was lacking in the theory that Darwin was keeping under wraps.

To Wallace's surprise his paper elicited neither hostile nor friendly responses in the scientific journals. Decades later Thomas Huxley said, "On reading it afresh, I have been astonished to recollect how little was the impression it made." A few scientists—other than Darwin, oddly enough—took Wallace's work very seriously. They were Edward Blyth, Sir Charles Lyell, and Joseph Hooker. Lyell even began keeping a notebook on the species question. He also went to see Darwin and urged him to hurry up and publish his ideas on evolution before another march was stolen upon him. Hooker chimed in with the same thought.

Concerning Wallace's paper Blyth wrote to Darwin from Calcutta, "Good! Upon the whole!" Wallace "put the matter well." Blyth asked, "Has it at all unsettled your ideas regarding the persistence of species?" Darwin had not told Blyth about his heterodoxy. As for Wallace's paper, Darwin called it "nothing very new . . . It seems all creation with him."

At the urging of Lyell and Hooker, Darwin started to write a monograph setting out his theory. But then the secretive scientist switched to writing a "big book on species." Was this just a case of nervous dithering? Or, being aware of previous writings, did Darwin fear that his idea would not be perceived as original?

A few years later, in 1858, Darwin, still working on his big book, received from Wallace a thick envelope, and this time the crypto-evolutionist was shocked to his toes. Inside Darwin found not only a letter but a *completed formal paper on evolution by natural selection*. What his friends had warned him about now had taken place. According to Darwin, the letter asked him to forward the paper to Sir Charles Lyell, and Wallace himself confirmed that fact in later years. Whatever else the letter said we do not know because it vanished, as did several other letters that Wallace had sent to the same recipient. This is odd. Darwin was so meticulous about saving correspondence that the publication of it, still going on, has run to fifteen volumes with approximately another fifteen volumes expected. The project has given rise to a "Darwin industry" among scholars.

Wallace wanted Lyell's opinion on his theory and, no doubt, if favorable, Lyell's help in getting it published by a scientific society. Wallace at this time did not belong to a scientific society. The *Annals and Magazine of Natural History* was a popular magazine, and probably the young man thought his new work would be taken more seriously if he could get it into a scientific journal. Lyell had a reputation for helping unknown naturalists, and Wallace knew from his correspondence with Darwin, that Lyell had taken an interest in his Sarawak Law. Wallace also might have expected Lyell to be sympathetic because the Welshman's theory of gradual evolution, conflicting with Genesis, was analogous to Lyell's theory of gradual geological change (uniformitarianism), which also conflicted with Genesis. But Wallace had no personal acquaintance with Lyell, one of the most distinguished scientists of his time (destined for burial at Westminster Abbey), and to Darwin, Lyell was a friend.

Darwin declared he "never saw a more striking coincidence." The contents of Wallace's paper, he wrote to the geologist, could serve as an abstract of the book he was writing. Wallace indeed had stolen a march, although he did not know it. Darwin had said nothing about his theory to Wallace, whom he pumped for information about tropical species. Darwin had kept secret his ideas. He corresponded with people all over the world, seeking information on plants and animals, and the flow of information "was almost always one-way." That was the observation of Janet Browne, associate editor of *The Correspondence of Charles Darwin* as well as a Darwin biographer.

Darwin forwarded Wallace's paper to Lyell with a despondent wail: "All my originality, whatever it may amount to, will be smashed." Without waiting for a reply, Darwin a week later wrote again saying, "There is nothing in Wallace's sketch which is not written out much fuller in my sketch, copied out in 1844" (not true). He asked Lyell to consult with Hooker, and this second letter closed with a promise not to "trouble" Lyell or Hooker on this subject again. Nevertheless, the second letter was followed by a *third* letter, which the writer called a "P.S." The final missive agonized, "It seems hard on me that I should thus be compelled to lose my priority of many years' standing, but I cannot feel at all sure that this alters the justice of the case." He added, "First impressions are generally right, and I at first thought it would be dishonorable in me now to publish."

In short, Darwin made known his intense desire to enjoy the priority of discovery, and he dumped the moral question on his two best friends, making no explicit recommendation. No doubt his cronies were thinking, *I told him so.*

Darwin always had supported Lyell's radical theory of geology, and now the geologist would not let down his old ally. Lyell and Hooker, "my two best & kindest friends" as Darwin flattered them, decided to present Wallace's paper to the Linnean Society along with two documents from Darwin. One of the two documents consisted of "extracts" from a "sketch" Darwin had finished in 1844 (not the 1844 document mentioned above); the other was an "abstract" from a letter he had written to Asa Gray in 1857.

In her biography Browne said that Darwin sent to Lyell and Hooker "an odd, mixed bundle . . . a very hasty culling of paperwork for a major turning point in biological science." According to Browne, Mrs. Hooker spent an afternoon copying extracts from Darwin's materials, "presumably extracts chosen by her husband," and Mrs. Hooker herself "silently" made some "helpful changes." Browne summed up the situation: "Darwin had little idea of what went forward until he saw the printed proofs several weeks after the event."

A great deal of revision was done. The 1844 "sketch" ran 230 pages (Darwin said in his autobiography), and the Hookers condensed it into three and a half printed pages of small size.

Lyell and Hooker presented the Darwin documents first and Wallace's formal paper second, and the materials were published by the society in that order. At the Linnean Society Darwin and his two friends had plenty of clout. Each of the three was a prominent scientist and each was a member of the Linnean Society Council. One member of the society withdrew his paper in order to make time for the materials offered by Lyell and Hooker.

As Browne commented, "No other venue could conceivably offer such obliging attention." And unless Lyell and Hooker moved quickly, Wallace might publish something first in a popular magazine. As for the Royal Society, it was not scheduled to meet soon, and moreover, getting something published by that august body would be complicated by referees, timetables, and committees. Would those deliberations help Darwin or Wallace? Darwin was embarrassed enough without having the Royal Society investigate what was going on. But if fairness was his objective, forwarding all the materials to the Royal Society would have been a better procedure.

Darwin did not attend the Linnean meeting. He stayed home mourning the death from scarlet fever of an infant son.

After the event Darwin thanked his friends and declared himself "*more* [emphasis Darwin's] than satisfied" with what taken place. But he wrote to Hooker that one thing surprised him: he thought his materials were to be only an "appendix" to Wallace's. That might have been more ethical as Darwin himself apparently thought, but on the record he never recommended it, before or after the Linnean meeting. Instead Darwin enjoyed the benefit of what his friends had done and put the responsibility on them. Lyell probably was used to that sort of thing. Before taking up geology he was a lawyer.

Compounding this Machiavellian mix, Darwin wrote to Wallace saying that he had "*absolutely* [emphasis Darwin's] nothing whatever to do in leading Lyell and Hooker to what they thought a fair course of action." In his autobiography Darwin went on in that vein, even asserting that he had "cared very little" about priority.

The naive young Welshman, eking out an existence in the jungle among cannibals, was delighted with what had taken place. "This," he wrote to his mother, "insures me the acquaintance of these eminent men on my return home." The cannibals were nicer to Wallace. When a thick, twelve-foot python coiled on top of his hut, they recruited a local snake expert and killed it for him. The reptile's skin ended up in London adorning a wall at the Linnean Society.

Wallace's response to Darwin was friendly and appreciative. "He must be an amiable man," remarked the English scientist.

In later communication Darwin fleshed out his story:

> You cannot imagine how I admire your spirit, in the manner in which you have taken all that was done about establishing our papers. I had actually written a letter to you, stating that I could *not* [emphasis Darwin's] publish anything before you had published. I had not sent that letter to the post when I received one from Lyell and Hooker, *urging* [emphasis Darwin's] me to send some MS, to them, and allow them to act as they thought fair and honorably to both of us. I did so.

Exhibiting the British talent for diplomatic language, Darwin, it would appear, did no more than make it possible for two distinguished scientists to act fairly and honorably.

Members of the Linnean Society were not much impressed by what was presented as the Darwin-Wallace theory. The London newspapers said nothing about it. The society's president, Thomas Bell, in his annual address to members lamented that 1858 had been a dull year scientifically. It, he said, was "not marked by any of the striking discoveries which at once revolutionize, so to speak, the department of science on which they bear."

Wallace's paper, entrusted to his friendly customer, physically disappeared, but its contents were preserved in the transactions of the Linnean Society. Darwin was able to revise and correct his work, even handing the job over to Mr. and Mrs. Hooker. Darwin informed the society that his 1844 work had not been intended for publication and therefore not written with care. With plenty of help, the rich, well-connected Englishman won out over the poor young Welshman.

A great variety of false and misleading statements have helped to cover up what took place in 1858. The Darwin exhibit at the American Museum of Natural History (not the online version) said only that Wallace "approached him [Darwin] with his own [unspecified] theory." That did not tell us much, besides being geographically impossible. Various textbooks and other works have reported that Wallace wrote only a letter and not also a paper, that Wallace wrote an "unfinished paper," that Darwin and Wallace were old friends, that the two prepared a paper jointly, that a "joint paper" was presented, that Wallace agreed in advance to the procedure at the Linnean Society, and, even that "by prearrangement Darwin and Wallace read their papers in London at the same meeting of the Linnean Society" (although neither man actually was present). In 2007 the British Museum of Natural History said online that the Linnean Society published a "co-authored paper."

By mistake or design much has been done, and still is being done, to cover up what took place in 1858. The science establishment does not want criticism of either Darwinism or Darwin.

It is my impression that textbooks treated Wallace a little better, at least for while, after a book on the 1858 affair, *A Delicate Arrangement*, was published in 1980. Author Arnold Brackman contended that Wallace was victimized by the "greatest conspiracy in the annals of science." Darwin seized the priority unfairly, Brackman argued, and might also have stolen from Wallace the "principle of divergence."

As for whether Darwin plagiarized anything in Wallace's paper, this depends in part on when the paper arrived at Darwin's house. The envelope with its postmarks disappeared. Brackman and at least one other researcher have thought it probable that the efficient Victorian mail system delivered the paper sooner than Darwin said it did. After reviewing the available evidence and somewhat suspicious circumstances, one can only express an opinion.

Concerning the matter of priority, in 1858 Darwin, compared to Wallace, did not present clearly or fully the theory for which the Englishman has been given credit. In London little attention was paid to the documents published by the Linnean Society, but this was not true in Dublin, where Wallace's paper drew fire from a geologist, the Reverend Professor S. H. Haughton. Haughton was shocked to see that, according to Wallace, evolution could progress "indefinitely from the original type." That, said the professor, was "at variance with all we know."

As for Darwin's contribution, Haughton objected only to the Englishman's "want of novelty." The geologist was a friend of Edward Blyth, with whom he corresponded, and that might explain the perceived lack of originality. Haughton summed up the Linnean presentations: "All that was new in them was false, and what was true was old." The only impressive aspect of the documents, said the Dubliner, was their presentation by Lyell and Hooker.

What did Darwin's presentation amount to? In the Linnean sketch of 1844 Darwin said that natural selection could produce a "marked effect," and he gave an hypothetical example. Suppose, he said, that a "canine animal" that preyed on rabbits found his territory becoming populated more by the hare and less by the rabbit. The hare (perhaps significantly, an example that was mentioned by Patrick Matthew) is a bigger and speedier animal that lives above ground. Rabbits burrow in the soil. In a thousand generations, Darwin said, the canine predator gradually would "adapt the form of the fox or dog to the catching of hares instead of rabbits." Darwin could see no more reason to doubt that possibility than to doubt the fact that "greyhounds can be improved by selection and careful breeding." With that analogy Darwin was claiming no more change than what is achieved by dog breeders.

In the letter to Asa Gray Darwin said that according to the "principle of divergence," new "species" could form as varieties of a species "seize on as many and as diverse places in the economy of nature as possible." This is why, said Darwin, a piece of land will support more life if occupied by a diversity of species than by just a few. The principle of divergence Darwin presented under its own heading separate from the principle of natural selection. Nevertheless, later in the same paragraph he said that the new variety or species would "exterminate its less well-fitted parent." That contradicted his claim that his principle of divergence brings about a greater number of species. And why would extermination result if each species finds its own place "in the economy of nature"? Darwin did not explain how the new species would form (Lamarckism?), and he gave no example. He admitted, "This sketch is *most* [emphasis Darwin's] imperfect; but in so short a space I cannot make it any better. Your imagination must fill up many wide blanks."

When Lyell had to lecture on evolution, he found Wallace's paper more useful than the two documents from Darwin. By 1867 four editions of *The Origin of Species* had come out, and Lyell was still complaining that Wallace's explanation was more lucid.

Wallace described a fierce competition for survival and said because of it "there is a tendency in nature to the continued progression of certain classes of varieties further and further from the original type—a progression to which there appears no reason to assign any definite limits." Surely it is stretch to say that the two theories were the same, and such a claim is surprising since the very title of Wallace's paper was, "On the Tendency of Varieties to Depart Indefinitely from the Original Type."

Wallace gave an example of evolution that involved a major change. Borrowing Lamarck's giraffe, he said that it evolved with a long neck in order to increase the available supply of food in time of scarcity.

In *The Origin of Species* Darwin copied Wallace by saying, "I can see no limit to the amount of change." In his final, 1872 edition Darwin also copied Wallace's giraffe example. After thirteen more years the author still was short on good examples from the wild. He derived his principles more from artificial breeding, while taking a keen interest in his collection of pigeons.

Darwin and Wallace thought much alike, but Wallace was ready to publish first and he also was willing to publish first. Darwin in twenty years did not manage to get his thoughts put together very well, and he was not willing to publish until his competitor forced him to do so.

As for the problem of divergence, has Darwin, Wallace, or anybody else really explained the diversity of life? Here is the basic problem: If new species replace the old species, how does the number of species increase? Remember, according to Darwin, evolution started off with just a "few forms" of life or one. Since then there have been billions.

Darwin said in *The Origin of Species*: "The extinctions of old forms is the almost inevitable consequence of the production of new forms." He also said, "extinction and natural selection go hand in hand." (Wallace too had taken that position.) Darwin meant that in the wild new, more competitive species replace the old, less competitive species—just as, in artificial selection, a breeder of pigeons disposes of unwanted birds so that they will not mate with the variety that the breeder is trying to encourage. And so—in theory—natural selection brings about a different life form just as does artificial selection.

It was Mayr's opinion that Darwin never did solve the divergence problem. He said Darwin "vacillated" among solutions to the problem of diversity, was "confused," "contradicted" himself, and ultimately "failed to solve the problem indicated by the title to his work." A Japanese professor, Soshichi Uchii of Kyoto University (one of Japan's finest), has said that he spent several years of "hard work" trying to figure out what was Darwin's principle of divergence and whether it had anything to do with natural selection.

Hence paleontologists have been known to quip, "The book called *The Origin of Species* is not really on that subject." This locution (using *species* in its plural sense) reportedly originated with the waggish, though very eminent, paleontologist George Gaylord Simpson.

The humorists also might have in mind something Darwin said in his autobiography: "Although in *The Origin of Species*, the derivation of any species is never discussed . . ." Darwin went on to explain that in the book he made an exception for the human species in saying "light would be thrown on the origin of man" so as not to be accused of hiding his views on that subject. The autobiographer overlooked his borrowed giraffe, which the textbook writers seized upon as their best example of natural selection.

Darwin knew that birds which made it to the Galapagos Islands from South America had diverged from their ancestral species. Thus isolation in a new environment could bring about change, though the Galapagos finches, thirteen closely related species, remained finches with not much physical difference among them. Darwin's finches found different sources of food such as insects, cactus pulp, seeds, and blood (the vampire finch attacks the booby bird), and so the beaks differed in size and shape. At least some of the finch "species," however, have been seen to mate with each other and produce fertile offspring.

Mayr, the dean of Darwinism (he died in 2005 at the age of one hundred) believed it rare for diversification to occur in the absence of geographical isolation. Yet isolation is much less easy to achieve than one might think. Geneticist Steve Jones, as reported in *The Third Culture*, wanted to find an isolated population of fruit flies and experiment with it. Jones traveled all over the deserts of California and into Mexico looking for isolated populations but without finding any. "After three years, a lot of money, and a deep tan," he said, "what we basically found was that those populations weren't isolated at all; there were flies flying in and out all the time."

If Mayr and Jones are right, how do the Darwinists explain hundreds of species of cichlid fish endemic to Africa's Lake Malawi? Darwin envisioned his "canine animal" on an island, and Huxley raised the usual lack of reproductive isolation as a major objection to natural selection. Mayr acknowledged the cichlid problem as "difficult" in his book *Systematics and the Origin of Species*. To account for cichlids' diversity, Lamarck would have cited their different behaviors and nature's self perfecting tendency.

The presentations of 1858 might have been forgotten entirely, or at least for a long time, if in the following year *The Origin of Species*, an "abstract" of the "big book on species," had not struck the world like a thunderstorm at the picnic. The book's accumulation of data and argumentation made a significant difference in the theory's reception. (Darwin himself called the book "one long argument.") Wallace, always generous, commented favorably on the book. He wrote to Darwin, "My paper would never have convinced anybody or been noticed as more than ingenious speculation . . . All the merit I claim is having been the means of inducing *you* [emphasis Wallace's] to write and publish at once."

Wallace's brother-in-law, Thomas Sims, took a different view. After *The Origin of Species* came out, Sims complained that the author did not mention the Sarawak Law and did not mention by name Wallace's 1858 paper. Wallace countered that Sims misunderstood the situation and that Darwin's conduct had been "disinterested." It was not until after Darwin died that Wallace found out what a turmoil he had caused. At that point he must have realized that of the three eminent men who held his fate in their hands, none was truly disinterested. Perhaps this is what caused Wallace to remark in his last book, *Moral Progress*, that in times of "great temptation" a person's true nature comes out.

Wallace always gave Darwin much credit for their theory, even writing a book titled *Darwinism*. Possibly conscience-stricken, Darwin once wrote to Wallace, "You are the only man I ever heard of who persistently does himself injustice and never demands justice." It was typical of Wallace that he rejected a membership invitation from the Royal Society, saying he was not worthy. The society then rejected his rejection, forcing an FRS upon him.

Upon his return to England Wallace continued to publish scientific work. As a founder of biogeography he demarcated the Wallace Line that separates the Asian Pacific islands from those geologically and biologically related to Australia. (It was Huxley who named the line.)

Unfortunately, Wallace's many publications, although excellent, on the whole did not make much money and the author, unlike Darwin, possessed so little skill at investing that he and his family lived in poverty. At the lowest point a woman came to Wallace's rescue. This was Miss Arabella Buckley, who had been a private secretary to the now deceased Sir Charles Lyell. According to Wallace's autobiography, Miss Buckley in 1880 was his "most intimate and confidential friend" and to her alone he confided his financial predicament. Miss Buckley on a visit to Darwin explained the situation, and he assigned to Huxley, his chief wire puller, the task of obtaining for Wallace a government pension. This was accomplished with the help of certain other scientists, and to Wallace it was a "joyful surprise" to learn that he would be receiving a pittance of 200 pounds per year.

In 1882 Wallace served as one of Darwin's pallbearers, marching with the two dukes and the American envoy, and one must hope that he had been able to afford a proper suit of clothes. Because fancy attire was needed, Wallace later pleaded poor health in order to avoid visiting Buckingham Palace, where King Edward VII wished to present him the Order of Merit. The medal had to be delivered to Wallace's home in Dorset by a colonel in the king's service. Next an artist arrived to draw a chalk portrait in color for the library at Windsor Castle.

Luckily, whatever was Wallace's investment in Darwin's funeral, it paid off. Fellow pallbearer James Russell Lowell invited him to come and lecture in the United States, and so the impecunious Welshman was able to tour that part of the New World, meeting famous people including President Grover Cleveland and visiting his expatriate brother in California. The tourist praised the superiority of American hotels and pronounced Harvard's Museum of Comparative Anatomy the best museum he had ever seen. Most of all, Wallace admired the mighty sequoia trees as a "living wonder" in itself worth a trip to America.

When Oxford University asked Wallace to attend the unveiling of a memorial to Darwin, the invitee suggested that Joseph Hooker might be a more appropriate choice.

As mentioned above, Darwin lost some confidence in natural selection and increased his respect for the effect of use and disuse, a concept proposed by his grandfather and Lamarck. Wallace lost some confidence in natural selection too, but he turned to spiritual causation.

Natural selection, according to both Wallace and Darwin, has the power to design a creature only *slightly* better than its competitors (just good enough to win the battle for survival), and Wallace pointed out that the human brain, with its great potential in music and mathematics, was far more powerful than what was needed for dwelling in the wild. (*In The Origin of Species* Darwin admitted that he had nothing to say about the origin of the "mental powers.") Wallace understood that very well from having seen the simple manner in which his Malay neighbors lived. Supporting him was the famous medical missionary Dr. David Livingstone, who said that in Africa he had seen no great struggle for existence.

That calls to mind a personal experience. As a small boy I fished in the Illinois River with moderate success. But when visiting a pristine lake in Minnesota, I was flabbergasted to haul in a dozen or so squirming prizes in just a short time. Such luck perhaps was not unusual in very ancient times. According to a Russian geologist, Prince Vladimir Kropotkin (1842-1921), pristine Siberia, despite its harsh climate, was full of wildlife. Invading Russians found the area, said Kropotkin, "so densely populated with deer, antelopes, squirrels, and other sociable animals, that the very conquest of Siberia was nothing but a hunting expedition which lasted for two hundred years."

As the most dramatic example of simple living, paleontologists now know that *Homo erectus* survived for about 1.7 million years carrying around a simple stone axe. Although the erectine's technology was not much, his longevity was phenomenal. (Will *Homo sapiens* do so well?)

Wallace decided that a "superior intelligence guided the development of man in a definite direction, and for a definite purpose, just as man guides the development of many animal and vegetable forms." Having escaped to a large extent the demands of natural selection, we humans now do a lot of selecting ourselves, usually by inadvertence. Beyond that, one might add, humankind now is in overall charge of the *planet*, and as we learn more about our ecological problems, this is a responsibility far more awesome than Wallace could have imagined.

Darwin's life of observation and theorizing somehow reduced drastically his enjoyment of poetry and the arts. The autobiography tells us that poetry and Shakespeare became "intolerably dull." The scientist's mind had become a "kind of machine for grinding general laws out of large collections of facts," and he did not know why this should have atrophied the part of the brain "on which the highest tastes depend." He did retain his enthusiasm for romantic fiction and also some appreciation for "fine scenery." Darwin mused that if he had to live his life again, he would read poetry and listen to music regularly so that the now atrophied part of his brain would be "kept active through use." Once more a note of Lamarckism.

Wallace went in the opposite direction. His appreciation of the arts greatly increased.

Charles Darwin as a young man was a believer in God headed for the priesthood, but his research led him to agnosticism. The opposite was the experience of Alfred Wallace. Once an unbeliever, Wallace followed a trail of data that, in his perception, led to a higher being. He even took to attending seances, as did many other prominent Victorians, such as the philosopher William James and the novelist Sir Arthur Conan Doyle.

Darwin, chronically ill, survived two months past his seventy-third birthday. The hearty Wallace lived to be almost ninety. Darwin did not shrink from the cross that he felt compelled to bear—the debilitating weight of a profound disenchantment. Wallace did not share that disenchantment, and although poor he lived a longer and happier life.

Chapter Two

The X Club Rules

There is indeed one belief that all true original Darwinians held in common, and that was their rejection of creationism, their rejection of special creation. Harvard University zoologist Ernst Mayr, *One Long Argument,* 1991

The Origin of Species sold well, but its theory was rejected by scientists and so the disappointed author came up with a strategy for success. To Huxley he said, "If we can once make a compact set of believers, we shall in time conquer."

Following publication of *The Origin of Species*, Huxley wrote a relatively favorable review. He warned Darwin that he had disagreements about how evolution worked, but showing that he was on evolution's side, Huxley gave this powerful assurance of help: "Depend upon it, you have earned the lasting gratitude of all thoughtful men. And as to the curs who will bark and yelp, you must recollect that some of your friends . . . may stand you in good stead. I am sharpening up my claws and beak in readiness."

There was the beginning of Darwinian militance. And intolerance.

82

Thomas Huxley was an extremely bright and articulate man who was born in poverty and spent most of his life struggling out of it. The self-effacing Wallace recalled his impression of Huxley:

> I was particularly struck with his wonderful power of making a difficult and rather complex subject perfectly intelligible and extremely interesting to persons who, like myself, were absolutely ignorant of [it] . . . I always looked up to Huxley as being immeasurably superior to myself in scientific knowledge, and supposed him to be much older than I was. Many years afterwards I was surprised to find that he was really younger.

Huxley's father was a schoolmaster who had a failing school, a wife, and eight children. Young Thomas escaped his miserable circumstances by devoting himself to intensive study. Foreshadowing his future career as Darwin's bulldog, he achieved some non-academic distinction by thrashing the school bully.

Huxley wanted to be an engineer but settled for a scholarship to medical school. There he showed great promise not only by winning prizes but by discovering a previously unknown feature of the human anatomy. This was a membrane next to the hair shaft, which still is known as "Huxley's layer." But merit was not enough to sustain his formal education. The precocious student's scholarship ran out, and so he took employment in the Royal Navy as an assistant surgeon aboard the HMS *Rattlesnake*. Like Darwin (but not *exactly* like Darwin, who was companion to the captain and the employer of a servant) Huxley studied exotic plants and animals while traveling aboard a warship.

Following his naval voyage Huxley wanted to publish the results of his research, but lacking connections, he could not find any source of funds. The wealthy Darwin for a similar purpose had been able to obtain a government grant of a thousand pounds by tapping Cambridge's old boy network. Darwin could have had a grant from the Linnean Society for writing *The Origin of Species*, but he turned that down. Eight years after the *Rattlesnake* voyage, the Royal Society sponsored publication of Huxley's research. Unfortunately, the delay made "sad havoc" of his formerly unique observations.

Four years after he had returned to land, Huxley at last found regular employment as professor of natural history and paleontology at the Royal School of Mines. The pay was not enough to get married on. The struggling scientist had a fiancee, Henrietta Heathorn, whom he had met in Sydney, Australia, and she would have to wait eight years for her sailor man to summon her. The prospective groom won a gold medal from the Royal Society, but sold it to pay a brother's debt. Eventually Huxley would supplement his salary with other appointments and also with his extraordinary skill as a writer and lecturer. The Royal Mines professor might have contributed much more to the advancement of science if, like Darwin, he did not have to eke out a living by writing and lecturing for popular consumption.

Miss Heathorn finally arrived in London from the antipodes, but in such poor health that a doctor gave her only six months to live. Her indomitable fiance vowed to marry her and cure her. That he did, and the two raised a large family.

Huxley wanted to make science a paying profession, and it irked him that while scientists labored on their own, the Anglican ministry was supported comfortably by the government. Evolution became his "Whitworth gun," an advanced firearm of the time, with which to fight the old order of aristocracy and Anglican privilege. The evolution controversy gave the freelancer a lot to write about, often working into the wee hours. An American paleontologist informed me that his colleagues in Britain, who often admire eloquence more than content, remember Huxley more respectfully than Darwin. Perhaps they also remember him for having helped to create their jobs.

Like Darwin Huxley suffered from illness and melancholia. He was a workaholic with a social conscience that kept him frenetically involved with numerous causes from public education to the state of the fishing industry. Withal, he found time to drill with a rifle in case of invasion by Napoleon III. In time the bulldog experienced breakdowns both physical and mental, and a "strain of madness" was perceived in him by a friend, the social reformer Beatrice Webb. "Ah! these great minds, seldom fit for everyday life," wrote Mrs. Webb in her diary.

In 1873 Huxley was worn out from overwork, deeply in debt, and his doctor said he needed three months of rest. At this time the poet Matthew Arnold encountered Huxley and was so affected by the scientist's haggard appearance that he left in tears. As in the case of Wallace, a woman came to the rescue. Sir Charles Lyell's wife saw that financial aid was urgently needed, and she took up the matter with Emma Darwin. Joseph Hooker's wife also joined the cause. The three women, evidently still encumbered with traditional values, persuaded Darwin and the X Clubbers to take up a collection. The men did so, and they also solicited funds from friends, including two industrialists, one of them being Sir Joseph Whitworth, the gun manufacturer.

Darwin then slipped into Huxley's bank account 2,100 pounds so that he could meet his financial obligations, get some treatment, and take a vacation. Nervously Darwin explained in a letter that the money came from him and some other friends. To the relief of all Huxley accepted the gift gracefully and gratefully. The vacation did not work out so well as had been hoped; the patient could not resist the opportunity to study geology and fossils in Europe. But he did come back looking refreshed.

Huxley, as mentioned earlier, disagreed with Darwin about both gradualism and natural selection. He wrote in an anonymous review of *The Origin of Species*:

> The combined investigations of another twenty years may, perhaps, enable naturalists to say whether the modifying causes and the selective power, which Mr. Darwin has satisfactorily shown to exist in Nature, are competent to produce all the effects he ascribes to them, or whether, on the other hand, he has been led to over-estimate the value of the principle of natural selection, as greatly as Lamarck over-estimated his *vera causa* of modification by exercise.

On the twenty-first anniversary of the publication of *The Origin of Species*, Huxley addressed a gathering of scientists and the socially prominent at the Royal Institution, and he disappointed the book's author by not mentioning natural selection, the concept of which Darwin was most proud.

Concerning natural selection Wallace and perhaps Darwin drew inspiration from Thomas Malthus, clergyman and mathematician, who was famous for having stated the gloomy principle: "The power of population is infinitely greater than the power in the earth to produce subsistence for man." (The tendency of population to grow geometrically the clergyman found, as he acknowledged, in the writings of Benjamin Franklin, who on that basis had recommended England's annexation of Canada after the French and Indian War.) Therefore, decided Malthus, inevitably there must be checks on population growth such as starvation, disease, and war. The industrial age with its more efficient agriculture was to prove Malthus wrong, at least temporarily, but concerning life in the wild his point was valid. In *The Origin of Species* Darwin calculated that in the case of elephants, a slow-breeding species, a single pair in five hundred years could have 15 million offspring if the process were not checked in some way.

As for gradualism Huxley could not see much of that in the fossil record. In 1894, long after Darwin's death, he wrote to the geneticist William Bateson:

> I see you are inclined to advocate the possibility of considerable "Saltus" on the part of Dame Nature in her variations. I always took the same view, much to Mr. Darwin's disgust, and we used often to debate it.

Saltus is the Latin word for a "jump." As one who disbelieved in Darwin's gradualism, Huxley was a saltationist.

Evolution by insensibly fine steps was important to Darwin because he could demonstrate that sort of change in the artificial breeding of plants and animals. Evolution, he argued, was an extension of the same process. Of course, it was a weakness of this analogy that the breeders of dogs and pigeons, whatever success they enjoy, end up with dogs and pigeons—as indeed the famous Galapagos finches, however much they varied, remained finches. Nevertheless, to depart from gradual transformation, said Darwin, was "to enter into the realms of miracle, and to leave those of Science."

As Darwin saw it, if the demands of the environment did not shape new species, what did? What else could be the source of design?

Huxley countered with the Ancon sheep. In Massachusetts there suddenly had appeared a breed of short-legged sheep that became popular with farmers because they could not jump fences. (Eventually the Ancons were abandoned because the Merino sheep provided better wool.) With that evidence Huxley argued that evolution somehow proceeded with big and sudden changes like the reactions that take place in organic chemistry. On Darwin's side it has been noted that the Ancon sheep represented a pathological condition, a failure of cartilage to form in the joints, and that the Ancons could not have survived in the wild.

Mario di Gregorio, who wrote the scholarly book *T. H. Huxley's Place in Natural Science*, counted four different versions of evolution that Huxley presented to different audiences. When lecturing to the Royal Society, the most sophisticated, he did not actually mention evolution, but he gave evidence for it. As for Huxley's own theory, di Gregorio could not find one. He concluded, "It is not possible to form a picture of what Huxley actually thought nature got up to in the production of the objects he studied."

Huxley and Darwin did agree that creationism was out. Some kind of purposeless (non-teleological) evolution was in. George Campbell, the science-minded eighth duke of Argyll, observed drily that looking for purpose was what Darwin did all the time and managed to find one in each feature of every organism.

As Darwin had recommended, Huxley undertook to form a compact set of believers. First he chose some representatives of different sciences and founded a dining society calling itself the "Thorough Club," which was dedicated to "the propagation of common honesty." That did not last long. Next, Huxley decided to recruit from his friends those who were evolutionists and not constrained by religious beliefs. The new group first got together on November 3, 1864, and after a while it took the name of "X Club." The club consisted of Huxley, seven other scientists, and the philosopher Herbert Spencer. Except for Spencer all were members of the Royal Society. The club's godfather, Charles Darwin, was welcome as a guest, but the recluse lived out of town and did not make much use of the privilege. The group decided on the name X Club because for long it could not decide on a name, and one of the wives, probably Mrs. George Busk, suggested X Club as appropriate for such a nameless organization. The club's only rule was to have no rules, except one against non-attendance.

According to Spencer's autobiography the club had no "avowed purpose beyond the periodic assembling of friends" and there took place a considerable amount of "badinage." Spencer admitted, however, that after dinner the members discussed scientific matters and the affairs of the scientific societies, especially those of the Royal Society.

Besides Huxley and Spencer, the group included the botanist Hooker, the anatomist George Busk, the chemist Edward Frankland, the mathematician Thomas Archer Hirst, the archaeologist and entomologist John Lubbock, the physicist John Tyndall, and the mathematician and philosopher William Spottiswoode. The compact band of evolutionists was adept at dealing with people and wielding influence. Huxley was a natural leader. Lubbock, who published an insightful essay on the subject of tact, was elected to the Parliament.

As the club got into gear, financial help came from British capitalists. They wanted to promote scientific and technical education in order to compete better with the rising industrial might of Germany. When it came to an arms race with Germany, Huxley gave that strong support too. Two wealthy allies contributed 3,000 pounds to help the poorer members of the Royal Society, and these funds were parceled out by Hooker. Huxley advised both Cambridge and Oxford universities on science endowments and faculty selection. In his book *Charles Darwin: The Man and His Influence* Peter Bowler offers a glimpse of the X Club

> Their different interests in evolutionism might have led them to dispute with one another in public, but instead they maintained a united front against the common enemy and worked tirelessly to insure that evolutionary papers would be published and that scientists favorable to their cause would have access to research funding and academic appointments.

Another look at the club is provided by the biography *Darwin* by Adrian Desmond and James Moore:

> [The group] constituted themselves into a sort of masonic Darwinian lodge invisible to outsiders: a dining club 'untrammeled' by any theology . . . Maneuvering inside the Royal Society, they altered the election procedures to get their allies elected and soon were pulling the Presidential strings.

Desmond and Moore said the club intended to "place an intellectual priesthood at the head of English culture."

Just a few weeks after the club's formation, the Royal Society awarded its highest honor, the Copley Medal, to Darwin. Hooker and Huxley each received one later. Some X Clubbers assisted the founding of periodicals: *The Natural History Review, The Reader,* and *Nature.* The first two failed. (In *The Reader* Huxley's Catholic-bashing did not help.) In 1869, *Nature,* published by the Macmillan Company, became the club's permanent outlet. Norman Lockyer, formerly of *The Reader's* science department, edited the new, wholly scientific journal. Huxley was a great admirer of the German natural philosopher and playwright Johann Wolfgang Goethe, and to lead off the first issue, Huxley assembled a series of Goethe aphorisms concerning nature. He also wrote a short article, and Lubbock contributed a book review. Within a year *Nature* published several articles favoring Darwinism, two of which were written by Darwin himself. "Far more than any other science journal of the period," according to Janet Browne, "*Nature* was conceived, born, and raised to serve polemic purpose."

The conspirators got Hooker elected president of the British Association for the Advancement of Science. They backed Lubbock when he stood for Parliament, and his election gave the group access to the prime minister. The club's MP was elevated from hereditary baronet to baron, becoming Lord Avebury.

A physicist member of the club jarred the whole Western World with a speech in Belfast. This was the Irishman John Tyndall, and at the time he was president of the British Association for the Advancement of Science. Tyndall averred, "We shall wrest from theology the entire domain of cosmological theory." In a grand paean to materialism, the physicist declared that one could find in matter "the promise and potency of every form and quality of life."

For about three decades the X Club functioned as a self-appointed government of Britain's scientific affairs. The secretive group, whose membership never increased, was active from 1864 until the early 1890s. It provided three presidents of the Royal Society and six presidents of the British Association for the Advancement of Science. From 1870 to 1878 Huxley, Hooker, and Spottiswoode all held offices in the Royal Society at the same time. At one point Hooker sat on fifteen committees of the Royal Society. Numerous offices in other organizations, such as the Chemical Society and the Mathematical Society, were filled by Huxley's *illuminati* or their selected candidates. Janet Browne noted, "One by one, the members infiltrated every government panel and committee that dealt with scientific affairs."

To those who took notice of the "Huxley set," its ringleader explained that their activities were purely social. But Huxley became known as a "wire puller," and the club got together just before every meeting of the Royal Society. The members had a simple tradecraft. The announcement of a meeting was sent by postcard and read as if it were the solution of a problem in algebra. For example, "X=15" would mean that a meeting was to be held on the fifteenth day of the month. The system could be humorous as well as "enigmatic" (Spencer's term). When the annual picnic was planned, the wives were referred to as the "YV's." Huxley was the "Xalted." (A three-way pun? Because of their shipboard years, Huxley and Darwin referred to each other as "well salted.") The long-winded Spencer was the "Xhaustive."

The guests of the club included two Harvard professors, the botanist Asa Gray and the historian John Fiske. Gray was more comfortable visiting the non-member Sir Charles Lyell, who like Gray acknowledged a Creator. Lyell never gave up his theism, and that "deeply disappointed" Darwin as he told Hooker. Gray held that a Creator was needed as a source of variations and design. Nonetheless, Darwin to one of his correspondents acknowledged the American botanist as the man who understood him best. "If ever I doubt what I mean myself," Darwin added, "I think I shall ask him!" As a visible manifestation of their extraordinary bond, the *alter ego* from Harvard grew a beard like Darwin's.

Professor Fiske helped arrange American tours for Spencer and Huxley, and he, rotund and witty, fitted in better than Gray with the irreverent X Clubbers. Memorably, the American entertained them with the story of how he had been warned by a cockney cab driver about that "'orrid hold hinfidel 'Uxley." When in England the professor was a frequent houseguest of the Huxleys. Fiske was much taken with the host's "eager burning intensity" and "clean cut mind." On his Sunday visits Fiske and the inventor of agnosticism did not go to church, but they sang psalms at the piano.

Fiske was thrilled by an opportunity to visit Darwin at his home. The Harvard historian later recalled the scientist's "quiet strength" and "guileless simplicity." Be that as it may, Darwin, by many contemporary accounts, had a courteous and engaging personality, which one can sense when reading his correspondence and even *The Origin of Species*.

In time word of the X Club's activities began to leak out, and according to Spencer, it was spoken of "with bated breath." Huxley reported to the group a conversation between two scientists that he overheard at the Athenaeum Club:

"I say," said one, "do you know anything about the X Club?"

"Well, they govern scientific affairs," came the response, "and really, on the whole, they don't do it badly."

The duke of Argyll held a different view. Although friendly to Darwin, he resented the Huxley set's extraordinary influence, and he complained about a Darwinian "reign of terror."

Like Patrick Matthew, the duke believed that beauty was an end in itself, and that it came from God. As evidence of divine creation, he cited esthetically pleasing examples of avian plumage, especially that of the argus pheasant and the iridescent male hummingbird. Darwin, he sniffed, found "a few *bits* of the truth." Argyll also disputed Lyell's uniformitarianism, saying that the Earth's contours had been shaped by cataclysms such as those recorded in the Bible.

Among friends Huxley demoted the nobleman to "dukelet."

Remarkably, Huxley and his coterie managed to dominate British science even though most British scientists were anti-Darwinian. Darwin and Huxley's primary strategy was to convert younger scientists, whom they persuaded that being Darwinian was both correct and advantageous. In 1868, Darwin wrote to the German physiologist William Preyer thanking the latter for his support, noting that German scientists were more receptive to his theory, and saying of his British colleagues:

> To the present day I am continually abused or treated
> with contempt by the writers of my own country; but the
> younger naturalists are almost all on my side, and sooner
> or later the public must follow those who make the subject
> their special study.

Compared to Darwin and his associates, Alfred Wallace felt that he was not much help in promoting evolutionism. He confided to a friend:

> I compare myself to a Guerrilla chief, very well for
> a skirmish or for a flank movement, & even able to
> sketch out the plan of a campaign, but reckless about
> communications & careless about Commissariat;—while
> Darwin is the great General, who can manouevre the
> largest army, & by attending to his lines of communication
> with an impregnable base of operations, & forgetting no
> detail of discipline, arms or supplies, leads on his forces
> to victory.

Like a certain general who allegedly avoided inspecting his troops for fear of what he might find, General Darwin had to accept his soldiers for what they were. And so being "Darwinian" did not mean that one accepted Darwin's theory. As Mayr has noted, it meant that one rejected creationism.

Herbert Spencer coined the term "survival of the fittest," but he later wrote a monograph titled, *The Inadequacy of Natural Selection* in which he strongly supported use and disuse as drivers of evolution. Conducting experiments with the human fingertips, Spencer discovered that they were far more sensitive than the skin elsewhere on the body when it came to distinguishing small objects. Such an ability would be helpful in sewing, he realized, but he had seen women sewing even with gloves on, and so he doubted that the extra sensitivity of the fingertips was needed for survival in the wild.

Mayr in his book *One Long Argument* classified Spencer as scientifically "neo-Lamarckian." Yet Spencer remained "Darwinian" because he believed in a purely secular evolution. Asa Gray, who possessed the most profound understanding of Darwin himself, believed in a theistic kind of evolution and therefore, despite the flowing white beard, was not a true Darwinian.

Since a Darwinian (or Darwinist) was a person who rejected creationism, and it did not matter if he rejected natural selection or gradualism, the X Club could be defined as an organization devoted to encouraging and propagating evolutionary theories that did not involve a Creator. In that sense Huxley's compact set of believers also could be described as a compact set of unbelievers striving to create more unbelievers. Adam Sedgwick had suspected an agenda like that. After reading *The Origin of Species* the Cambridge professor said of Darwin's theory:

> It is a system embracing all living nature, vegetable and animal; yet contradicting—point blank—the vast treasure of facts that the Author of Nature has, during the past two or three thousand years, revealed to our senses. And why is this done? For no other solid reason, I am sure, except to make us independent of a Creator.

The "X" rewarded its allies, and it could punish its opponents, as when Huxley and Hooker blackballed an anti-Darwinian candidate for the Athenaeum Club. The victim was the anatomist St. George Jackson Mivart. Darwin and his friends had tried to recruit Mivart as an ally, but he was a Catholic convert who insisted on finding some kind of spiritual element in evolution, and he wrote articles critical of Darwinian theory. In reference to gradualism Mivart asked, what good is half a wing? Natural selection he anathematized. And how could physiological variations be random when animals so unrelated as the squids and the vertebrates have similar eyes? Working by indirection, as he often did, Darwin took an American magazine article critical of Mivart and had it published in Britain as a pamphlet. (This was prior to the fracas with Samuel Butler, in which Darwin sponsored a publication attacking the novelist.)

The feud with Mivart came to a head when Darwin's son George published an article recommending some liberalizing of divorce laws in the interests of eugenics. Mivart slammed the idea as just the sort of moral breakdown to which natural selection theory was bound to lead. The anatomist wished to maintain friendly relations with Darwin, but the latter, complaining that Mivart distorted his and George's views, sent him a note breaking off all communication.

That excommunication was followed by another. The Catholic Church expelled Mivart, the champion of anti-Darwinism, because of his belief in evolution.

In its heyday the X Club was powerful and it could be swift, while attending carefully to decorum. The Westminster Abbey interment of Darwin is a good example. The first step was to obtain the family's permission, and it was decided that a request coming from the Royal Society would help with that. Darwin's cousin Francis Galton, who had suggested the abbey interment, was a member of the Royal Society and so as the most appropriate emissary, he was assigned to ask the society's president to seek the agreement of the Darwin family. This was a formality. The society's president was Spottiswoode, and he straightaway telegraphed the surviving Darwins. Eldest son William was strongly in favor of the abbey, and he persuaded his mother, Emma, who, however, refused to attend. Huxley and Spottiswoode went to see the Reverend Frederic Farrar, canon of Westminster. Farrar was science-minded and Darwin had praised his work on the origin of language. The two visitors talked the canon into to taking up their request with the dean of Westminster, the Reverend George Granville Bradley. Bradley was interested in science, too, and besides that, Huxley and Spottiswoode had supported his election to the Athenaeum. Now the dean, too, was in the bag.

Lubbock was a member of Parliament and also president of the Linnean Society. He got up a petition among his fellow parliamentarians for the abbey burial, and it was signed by several of the most influential. Meanwhile, members of the press were approached, and no doubt partly for that reason the newspapers praised Darwin to the skies.

In the *British Journal for the History of Science* Ruth Barton has provided this perspective:

> Darwin's burial in Westminster Abbey was an X-Club achievement. If Tyndall had not fought for Spottiswoode's nomination as Treasurer [of the Royal Society] in 1870, if all the 'X' had not put Hooker forward in 1873, and if Hooker and Huxley had not manoeuvered Stokes out of the Presidency in 1878, who can say who would have been President in 1882? But if Stokes had been President, Darwin would not lie in Westminister Abbey.

The funeral was a star-studded event with the choir singing, "Happy is the man that findeth wisdom and getteth understanding" (from Proverbs 3), a hymn commissioned for the occasion. The pallbearers were the duke of Argyll, the duke of Devonshire, the earl of Derby, James Russell Lowell, William Spottiswoode, Joseph Hooker, Alfred R. Wallace, Thomas H. Huxley, and John Lubbock.

In the middle of this solemn and elegant ceremony there emerged a bit of Darwinian eccentricity. As Emma had predicted, it was cold in the abbey and so the chief mourner, son William, "put his black gloves to balance on top of his skull, and sat like that all through the service with the eyes of the nation upon him," said Gwen Raverat in her memoir.

Huxley arranged the interment at warp speed, but it took him six years to write Darwin's obituary for the Royal Society. He must have struggled about what to say. Most notably Huxley, in sharp disagreement with Darwin, said in the obituary that it made no difference whether one believed in gradualism or sudden changes.

The X Club stayed behind the scenes as long as it was alive and even afterward. The grand wizard of the Darwinian lodge, Thomas Huxley, did not mention the subject in his autobiography, although for thirty years it was one of his major activities. Spencer's autobiography with its brief account of the club was published posthumously after the author died in 1903. Huxley and Darwin had passed away earlier, and the "X," as they called it, for long had ceased to exist and no longer was a matter of interest.

As Charles Darwin, the *eminence grise,* had planned, the acceptance of evolutionary ideas was aided greatly by Huxley, the X Club, and the club's carefully cultivated allies. Secular evolutionism prevailed in one guise or another. It appeared to offer a great deal of explanatory power, it made way for social change, and it put humankind in charge of its fate, undermining what many regarded as the stifling authority of the church. One might say the same of two other nineteenth century theories, Marxism and Freudianism. Darwinism, Marxism, and Freudianism became ideologies which, to their followers, seemed to explain everything; and all three ideologies broke down constraints that had been holding back powerful human desires.

"When a prisoner sees the door of his dungeon open," said George Bernard Shaw, "he dashes for it without stopping to think where he shall get his dinner outside."

Many of those ideologues have dashed from one prison into another: endless psychotherapy, totalitarian government, or in the case of the Darwinists, deterministic ultra-Darwinism. The last tells us that the gene, not the organism, is what counts; the gene is the ultimate unit of selection. This means that evolution basically serves the purpose of the genes, and *Homo sapiens,* therefore, is a robotic vehicle guided by circumstances and its tyrannical genome. One can find that written up by the ultra-Darwinian geneticist Richard Dawkins in his book *The Selfish Gene.* The ultra-Darwinian Daniel Dennett has commented, "Usually, other things being equal, what's good for the gene is good for the organism—and thus, you might say, for the species. But the gene is in the driver's seat."

Dawkins, by the way, does not assert that we humans evolved from the apes. He says we *are* apes.

Nineteenth century thinkers launched a powerful, three-pronged attack on the concepts of God and free will. If alive today, that letter writer in Manchester might ask, "What will happen if we teach our children that they are robots?"

Chapter Three

The Great Scopes Scam

To prevent his loyalty from delaying the trial I went to see the youngster and told him to go ahead and testify to what he had been told to say. John T. Scopes, *Center of the Storm*, 1967

Ask media savvy people what was the greatest propaganda film ever made, and they probably will tell you it was Leni Riefenstahl's *Triumph of the Will*, a documentary about a 1934 Nazi rally in Nuremberg, Germany. That film was great in cinematic terms, but owing to the fall of the Third Reich, its effect was short lived. On both stage and screen we Americans have pulled off a much bigger propaganda feat. For half a century our *Inherit the Wind* blitzkrieg, the stage play and films, has been persuading the world that anybody who questions Darwin must be a religious ignoramus who refuses to look at scientific facts. People assume that the *Inherit the Wind* productions are a fairly close rendition of the 1925 trial in which John T. Scopes was convicted of breaking the law by teaching evolution, but the drama departs from the facts in grossly misleading and slanderous ways, and one detractor has called it "Inherently Wind." Yet the play ran on

Broadway for three years with a record-breaking 805 performances, and that persuaded the *Encylopaedia Britannica* to take up the trial, the most famous in American history.

The truth about the "Scopes Monkey Trial" is far more interesting than the fiction. For example, the play fails to mention that the top scientist helping the defense did not believe in Darwinism himself and published a theory of his own. This was Henry Fairfield Osborn, president of the American Museum of Natural History. Osborn called his theory "aristogenesis" and he termed it "contra-Darwinian." The biochemist Roger Lewin has described Osborn's theory as a "very aristocratic view of the world," and somebody has commented that aristogenesis was just the sort of theory that one might expect from a man who owned a castle-like mansion situated on a high ridge overlooking the Hudson River.

Osborn, the nation's most influential paleontologist, evidently did not care what exactly was being taught in the schools. He just wanted the children to learn *some* kind of evolutionary theory, right or wrong. Even the ultra-Darwinian Richard Dawkins does not necessarily stand by what the schools are teaching. In the *New York Times* article mentioned earlier he said, "It is still (just) possible for a biologists to doubt" the importance of natural selection, and he noted that "a few claim to." Such diversity I have yet to see in the textbooks.

The stage production of *Inherit the Wind* was written by Jerome Lawrence (originally Jerome Schwartz) and Robert E. Lee, who said the Scopes trial was the "genesis" of their work. The play opened in 1955. It spawned three motion pictures, and the play itself still is performed from time to time. As an indication of the play's continuing popularity, the libraries in my county stock from one to three copies of the script in book form. In 2005 a Darwin festival in England made the evolution drama the "centerpiece" of the event. Creationists complain that *Inherit the Wind* films often have been used by American schools as instructional material for classes in science, history, and social studies. If so, that is a case of malpractice in pedagogy.

The motion picture referred to here is the one that features Spencer Tracy, who won an Academy Award for his performance as the character based on the defense attorney Clarence Darrow; Fredric March, who won an award in Germany for his performance as the character representing prosecuting attorney William Jennings Bryan; and Gene Kelly, better known as a dancer, who somehow was cast as the character representing the cynical, pro-Darwinian columnist H. L. Mencken.

Both Bryan and Darrow were known for their outstanding oratorical abilities. Darrow was a labor lawyer and criminal lawyer. Bryan, a silver-tongued populist known as the Great Commoner, three times had run for president as the candidate of the Democratic Party. He also had served as secretary of state, resigning in 1915 from the administration of President Woodrow Wilson. Bryan quit the cabinet because of his belief that Wilson planned to maneuver the United States into World War I, and two years later Wilson did obtain from the Congress a declaration of war. Bryan was not the clueless bumbler depicted by *Inherit the Wind*.

In Lawrence and Lee's play angry residents of a small town fire an idealistic young high school teacher for breaking a law against teaching evolution, and the teacher is put in jail to await trial. Hollywood adds to the story people throwing rocks at the prisoner, one hitting him in the face, and there is an implied threat of lynching. The truth is, the teacher, John T. Scopes, was a popular figure in town, he was not jailed, he was not fired, nobody threw rocks at him, and this hero of American mythology did not actually teach evolution. Furthermore, his alleged idealism was tainted when Scopes, in order to help cook up a trial, persuaded his students to commit perjury by testifying that he *had* taught evolution even though they could not recall such a lesson. Scopes did not remember teaching the subject either. He avoided perjuring himself by pleading innocent and refusing to take the stand.

How to make sense of all this? We must go further back in time.

In the early decades of the twentieth century, evolutionary biology obviously was in chaos. As Ernst Mayr has said, the "field was split into several camps of specialists furiously feuding with one another." There was no X Club to maintain a united front, natural selection was not well accepted, gradualism lacked for evidence, Mendel's laws of inheritance contradicted Darwin's notion of blending inheritance, and the newly discovered phenomenon of mutation had given rise to mutationism (also known as Mendelism) as a competing theory of evolution. (Huxley would have been pleased). Geneticists took to inducing mutations in fruit flies, expecting some new species of possibly improved fly. What they got (thankfully) were a lot of sick, dead, and weird-looking flies that could not survive outside the laboratory.

The previous chapter mentioned William Bateson, one of Thomas Huxley's fellow saltationists. Bateson was the geneticist who coined the word *genetics,* and in 1921 he was very frustrated by the shambles that evolutionary theory had fallen into. He spoke about that at some length to the American Association for the Advancement of Science, which was meeting in Toronto, Canada. The speech was titled "Evolutionary Faith and Modern Doubts." Concerning natural selection Bateson said it was a "plausible account of evolution in broad outline, but failed in application to specific differences."

In other words, it was a good theory until one examined the evidence.

The geneticist summed up: "When students of other sciences ask us what is now currently believed about the origin of species, we have no clear answer to give." As if that were not enough to stir up trouble, Bateson added, "Faith has given place to agnosticism." No doubt the public was surprised to learn that faith had been such an important part of evolutionary theory. Following Bateson's speech "The more noisy newspapers went off in fully cry, with scare-headings, 'Darwin downed,' and the like," the geneticist complained; and this hullabaloo produced in the United States a revolt against evolution itself.

In *Nature* Bateson admitted it was he who "all unwittingly dropped the spark which started the fire." The British geneticist expressed the opinion that perhaps people in "Kentucky" and "'Main Street'" (a reference to Sinclair Lewis's novel about a narrow-minded small town) did not really need to know about evolutionary theory, anyway. Those people might be better off, he said, if they studied the Rock of Ages instead of the age of rocks, just as the anti-evolution politician William Jennings Bryan was urging them to do.

In line with this prominent scientist's suggestion, bills to prohibit the teaching of evolution were introduced in some state legislatures. One in Tennessee was called informally the "Monkey Bill" because it was the general understanding that evolution meant that man had evolved from the monkey (not the ape). This version of evolution did not result from ignorance on the part of the people of Tennessee. It came from Darwin's book *The Descent of Man,* which in fact stated that man evolved from the monkey.

In 1925 the Monkey Bill became law as the Butler Act. Not a great deal of attention was paid by the press in Tennessee, but the event caused a worldwide stir among intellectuals. Among those condemning the law were Albert Einstein and even the anti-Darwinian George Bernard Shaw. Shaw loathed natural selection, but he perceived the alternative to evolution as "the monstrous nonsense of Fundamentalism."

Looking at the other side of the argument, if a scientist of Bateson's stature did not know where species came from and if there were no generally accepted theory, why should socially disruptive ideas be taught in school? Surely, the proper procedure would have been to tell the truth, which was this: in geological history a vast number of different species have come and gone for causes not known although various explanations have been proposed.

Not only was there no well established theory of evolution, but the public worried about the moral problem raised by that letter writer in Manchester, England. How could morals be taught when in nature progress came not from God's love but natural selection, which is to say, starvation and brutality?

Just the year before the Butler Act passed, a shocking instance of moral depravity had unsettled the whole country. In Chicago Nathan Leopold and William Loeb murdered fourteen-year-old Bobby Franks in what the press called a "thrill killing" and a "motiveless murder." The killers were not poor and ignorant. To the contrary, they came from wealthy families, and they were brilliant students. Leopold, only nineteen, already was in law school. Loeb, eighteen, was the youngest graduate in the history of the University of Michigan. These academically accomplished teenagers were familiar with the "God is dead" philosophy of the Darwin-influenced Frederic Nietzsche, and they believed themselves Nietzschean supermen who could commit the perfect crime.

Unfortunately for Leopold and Loeb, there was nothing superior about their murdering ability. They left Leopold's prescription glasses with its unusually hinged frames near the body, and they failed to dispose of the typewriter he used to write a ransom note. Neither did it help that they left the body in an area which Leopold was known to frequent as a birdwatcher. The two Nietzschean supermen were book smart but not street smart.

The Leopold and Loeb families hired Clarence Darrow to defend their sons. Known for his rumpled suits and colored suspenders, Darrow was disarmingly folksy and very effective. But nobody expected the boys to get off. They had confessed, even bragged about what they had done, and as Leopold explained to the police, "It was just an experiment. It is as easy for us to justify as an entomologist in impaling a beetle on a pin." The public was horrified. Chicagoans were used to bootleggers killing each other for profit. They were not used to privileged and highly educated young men murdering children just for their amusement.

Loeb's mother would not believe he had committed the crime. "I told her it was true, but she wouldn't believe me," complained the young killer. "What hurts is that she won't believe me." Until a police captain advised him otherwise, this immature teenager thought he would spend only a few years in jail, then come out, "work hard," and "amount to something."

The parents asked of Darrow only to avoid the death penalty. That he did by waiving a jury trial and concentrating his persuasion on just one person, the judge. Darrow made a twelve-hour plea, over two days, in which he wept and said it was not fair to hang these studious young men for believing in a philosophy they had been taught in college. Darwin's mechanical theory of evolution supported Darrow's philosophy of determinism, and the lawyer portrayed the defendants as pitiful victims of nature, their actions determined by heredity and environment. They should no more be executed for killing Bobby Franks, Darrow argued, than an old sow ought to be tried in court and executed for lying down on her piglets and causing their death.

The defense achieved its objective. The teenagers were convicted, but instead of the death penalty sought by the prosecutor they were given life plus ninety-nine years. As a possibly a major factor in the case, the prosecutor angered the judge by insinuating that he was friendly to the defense. Darrow collected less than half the fee that had been planned. But to Darrow the publicity generated by the "Crime of the Century" possibly made up for the shortfall.

Loeb was killed in prison. Leopold was paroled in 1958.

Many observers thought the Butler Act would never be enforced, and among them was the governor of Tennessee, Austin Peay. Upon signing the measure Peay commented that the schoolbooks in his state did not teach evolution. He was wrong. Many did take up the subject although evidently without causing much trouble if the governor was not aware of the books' contents. One such textbook was *A Civic Biology* by George William Hunter, and this was the book that figured in the trial of Scopes. The book had been published in 1914 when the author was head of the biology department at De Witt Clinton High School in New York City. At the time of the Scopes trial Hunter was a professor of biology at Knox College in Galesburg, Illinois, which today has the biology wing of a building named for him. The wing commemorates the professor and his textbook's "small but important role in the history of modern science."

A Civic Biology was tiny in size compared to the massive, many-pictured biology books that high school students lug around today, but I would guess that Hunter's economical book offered as much information as the youngsters could absorb. Contained was a wealth of practical information and illustrative experiments.

Compared to Bateson's speech about evolution, however, the 1914 textbook was out of date and unjustifiably dogmatic about what it called the "doctrine of evolution." Hunter taught that according to "the great Englishman Charles Darwin," small variations in heredity gradually bring about new species as the result of natural selection. New species also arise suddenly by mutation, said Hunter, crediting that discovery to the Dutch scientist Hugo de Vries. Politically correct for its time, *A Civic Biology* said that the highest achievement of evolution was the civilized Caucasian race of human beings inhabiting Europe and America. (There was Darwin's most Favoured Race.)

Considering what Bateson had said and the fact that evolution by large mutations had been rejected, the Tennessee legislature was more up to date than the world's scientists. The latter were urging that Hunter's book be taught to the children of Tennessee as the established truth while the state's lawmakers were perceiving it, correctly, as flawed.

Left to themselves the people of Tennessee and their governor might have forgotten about the Butler Act, but it was not up to them. In New York City the American Civil Liberties Union announced that it would defray expenses for a test case, and a New Yorker of evolutionist bent happened to be residing in the little town of Dayton, Tennessee. This was George Washington Rapplyea. He had come to Dayton to manage a formerly prosperous coal mining and steel-making company that was selling off its assets. In the ACLU offer Rapplyea saw an opportunity to stimulate business and put little Dayton on the map. It was a matter of local concern that the population of the town had dwindled from 3,000 to 1,800, chiefly because the coal and steel operation had closed down.

Rapplyea took his proposal to Fred Robinson's drugstore, a major Dayton social center (another, less prestigious hangout was the barber shop). In those days people often took refreshments at drugstores. One could sit at a counter or a table and enjoy ice cream or a soft drink. "Doc" Robinson, the drugstore proprietor, happened to be chairman of the town's school board. Also present that day, sitting at a (soon to become historic) round table, was George White, superintendent of county schools, and Sue Hicks, a city attorney. Hicks was male but his mother died in giving him birth, and the grieving father gave his son her name, which for a country lawyer was unforgettable.

Over soft drinks the group was discussing teacher selection for the high school's next academic year and also the Butler Act. Hicks thought the new law would have no effect unless tested in court. Rapplyea, having studied geology to become a mining engineer, was an evolutionist. Hicks was undecided about that for lack of proof. White opposed evolution on the grounds that it undermined religious faith. As for Robinson, I have found no opinion expressed by him, and the stage drama might be correct in saying that the druggist never had opinions since they were bad for business. The four drugstore loungers made a pretty good focus group.

Rapplyea sat down and suggested that he and the men present organize a test case for the ACLU, where, he intimated, he had some contacts. The case, he said, would surely attract national attention and bring a lot of visitors to Dayton. The lawyer and the druggist liked Rapplyea's idea. There was talk of getting famous people involved, such as the novelist H. G. Wells, who had written *The Time Machine* and *The War of the Worlds*. Superintendent White observed that the school year had just ended, but that point was considered minor. Other citizens dropped by, and they also favored the project.

What did Dayton have to lose? All the Gang of Four needed now was a defendant.

Dayton happened to have a young man who appeared to be the perfect defendant, a twenty-four-year-old high school teacher by the name of John T. Scopes. He was single, popular, and self-effacing—obviously not a publicity-seeker. The *New York Times* would describe him as "lanky and grave-eyed." But would Scopes sign on, and had he ever taught evolution? Scopes had planned to leave town for a summer job selling cars, but luckily he had delayed his departure with the expectation of dating a certain beautiful blonde.

Inherit the Wind invents for the defendant a distraught fiancee, and the Bryan character, demanding that she reveal Scopes's ideas about evolution, torments her on the witness stand to the point of exhaustion. In doing so the interrogator breaks a confidence; prior to the trial he had inveigled the young woman into telling him about her fiancé's ideas. *Inherit the Wind's* frail beauty also is treated sadistically by her father, a Christian minister, of whom she has been terrified all her life. The fictitious father, with daughter present, denounces the Scopes character publicly and calls upon God to strike him down.

When they wanted to be inflammatory, Lawrence and Lee did not hold back.

Scopes, who actually had no fiancee, was summoned to the drugstore from playing tennis. Perspiring heavily, he was given a cold drink and a chair. Rapplyea played his quarry with care. Said he: "John, we've been arguing, and I said that nobody could teach biology without teaching evolution."

"That's right," said Scopes.

Doc Robinson's drugstore also sold textbooks (the proprietor called himself "The Hustling Druggist"), and Rapplyea showed

Scopes a copy of *A Civic Biology.* He asked, "You have been teaching 'em this book?"

"Yes," said the young man, explaining that he had used it for review purposes while filling in for the regular teacher, who had fallen ill.

"Then you've been violating the law," said Doc Robinson.

"So has every other teacher then," said Scopes. "There's our text provided by the state. I don't see how a teacher can teach biology without teaching evolution."

Robinson showed Scopes a newspaper story about the ACLU offer and asked whether he would be willing to let his name be used in a test case.

The teacher was facing the school superintendent, the chairman of the school board, a city attorney, and a prominent businessman. He needed a job to save money and pay for graduate school. What to do? Scopes waffled. He said, "If you can prove that I've taught evolution, and that I qualify as a defendant, then I'll be willing to stand trial."

"You filled in as biology teacher, didn't you?"

Scopes agreed.

"Well," said Robinson, "you taught biology then. Didn't you cover evolution?"

"We reviewed for final exams, as best I remember."

The vague reply did not faze the trial planners. They "apparently weren't concerned about this technicality," Scopes said in his memoir *Center of the Storm*.

The drugstore schemers and Dayton's need for business prevailed. Scopes then was arrested, after he himself fetched a deputy sheriff for the purpose. (Hollywood's version of *Inherit the Wind* prefers a dramatic classroom arrest in front of Scopes's students. The play starts with Scopes already arrested and in jail.) After the legal formalities Scopes went back to playing tennis, and in the subsequent rush of events he never did get to date the blonde.

Rapplyea telephoned the ACLU, which offered to cover the expenses for both sides of the trial. The prosecution turned down the financial help. Superintendent White gave the story to the *Chattanooga News* while Doc Robinson informed the *Nashville Banner* and the *Chattanooga Times*. There were no flies on this drugstore gang.

Dayton had been quicker on the draw than some much bigger Tennessee communities. Disappointed, they complained that Dayton was indulging in a "publicity stunt" and, anyway, the town was too small to host an event of national importance. Yet little Dayton was no stranger to fame. A local hero was Sergeant Alvin York, the most famous American infantryman of World War I.

Already we can say that, contrary to its reputation, the Scopes Monkey Trial did not come about because local Christians were angry about the teaching of evolution. The event was organized by a few town boosters, and the prime mover was George Rapplyea, a Yankee evolutionist. While seeking fun and profit, a few small town plotters were going to fool the world's sophisticated media into thinking that evolution was a burning issue in their community.

A special session of the grand jury was called in order to indict Scopes before some other town got ahead of Dayton. But what about evidence? Seven students testified against the teacher. Concerning how they were recruited, I have no specific information, but all were boys and perhaps it is relevant that Scopes was the high school's only athletic coach. Some of the witnesses disappeared at the last minute, and the coach had to find and retrieve them. Reportedly, he also advised the boys on what to say. As corroboration for that, in *Center of the Storm* Scopes admitted that the boys could not recall what they heard in the class held three months before. He also said that he himself "didn't remember teaching it [evolution]."

National attention was assured when William Jennings Bryan announced his willingness to help with the prosecution and accepted Sue Hicks's invitation to do so. Bryan for long had been a foe of Darwinism, which he asserted was an "unproven theory" and a detriment to the public morals. Meeting Scopes, he said, "You have no idea what a black and brutal thing this evolution is."

Bryan, of course, was familiar with Darwinism although both the play and the movie portray him otherwise. In arguing against evolution Bryan asserted that Darwin never explained how any particular species evolved into another (apparently he had read the autobiography). At the trial Bryan made another important point saying, "One trouble about 'evolution' is that it has been used in so many different ways that people are confused about it." Many years later, Ernst Mayr put that more strongly saying, "Since the 1860s no two authors have used the word 'Darwinism' in exactly the same way."

The legal superstar Clarence Darrow offered to help the defense. Encouraging Darrow's participation was the nationally prominent journalist H. L. Mencken. He told the attorney, "Nobody gives a damn about that yap school teacher. The thing to do is to make a fool out of Bryan."

While the ACLU's avowed aim was to preserve freedom of speech, Darrow, as he later wrote, had a different plan. He wanted to "focus the attention of the country on the programme [*sic*] of Mr. Bryan and the other fundamentalists in America." Looking forward to what he expected to be the greatest trial since the one held before Pontius Pilate, Darrow went from what the press called the "Crime of the Century" to one the press called the "Trial of the Century."

The ACLU did not want Darrow involved. The civil rights organization just wanted a friendly test case that would not get the defendant fired and could be appealed on constitutional grounds. The flamboyant Darrow, the ACLU knew, would confuse the issue and arouse fierce animosities. But Scopes wanted Darrow as a good "down-the-mud" fighter and so did the Dayton boosters. The boosters were so desirous of publicity that they even staged at the barber shop a fake shooting incident (using blanks), which ostensibly resulted from an argument about the Bible. Scopes complained to Rapplyea about that and no more such stunts took place.

Why was Rapplyea suspect? Scopes did not say in his memoir, but the coal executive was known to be a rough and ready sort of character. As a boy he had sold newspapers in New York, and he claimed to have fought every day to keep his corner.

The defendant went to New York and met with the ACLU. While there he also consulted with Henry F. Osborn. Darrow too met with Osborn and received advice on the subject of evolution although the attorney was well read on the subject. An evolutionist and an agnostic, Darrow had read not only Darwin but in addition the works of X Club members Huxley, Tyndall, and Spencer.

Osborn wrote a *New York Times* article in support of Scopes and at his request Major Leonard Darwin, a son of Charles Darwin, sent to the defendant a letter of encouragement.

At that time Major Darwin was president of the Eugenics Education Society, which sought to prevent "reproduction of the unfit," and in the United States Osborn was a leading eugenicist. Previously Major Darwin had served in army intelligence and had been elected to Parliament. His wife Charlotte was a second cousin, the daughter (not surprisingly) of a Wedgwood. In the 1920s and 1930s many thousands of feeble-minded and mentally ill people were forcibly sterilized in Europe and in the United States. In this country the decision to sterilize a retarded woman was upheld by the Supreme Court in 1927. The majority opinion was written by the liberal Justice Oliver Wendell Holmes, who said in reference to the woman's family, "Three generations of imbeciles are enough."

Today the practice of eugenics is condemned ethically, but an extreme form of it continues in North Korea. According to a United Nations report of 2006, North Korea maintains prison camps for dwarves and other persons considered genetically unfit. The imprisoned are allowed to marry but not to reproduce. They are given hard labor and used for scientific experiments.

Osborn did not go to Dayton as one of the scientific advisers on evolution, explaining to Scopes that his wife was ill. If Osborn had gone to Dayton, there could have been embarrassing questions such as, "Whatever happened to Nebraska Man?" In 1922 Osborn had identified a tooth fossil from Nebraska as that of an "anthropoid ape" which the press called "Nebraska Man," humankind's only ancestor discovered in America. Perhaps giddy with this great accomplishment, Osborn used it to make fun of the anti-evolution Bryan, who had been elected to Congress from Nebraska. Bryan, said the paleontologist, was the "most distinguished primate which the State of Nebraska has thus far produced." (Actually, Bryan had been born in Salem, Illinois.) Bryan fired back, "Professor Osborn is so biased in favor of a brute ancestry . . . that he exultantly accepts as proof the most absurd stories."

Bryan happened to be right about Nebraska Man. By the time the Dayton trial came along, Osborn had stopped talking about his great discovery, and he did not mention it in his *New York Times* article. Nine scientists had raised questions about the tooth, and in 1925 one of Osborn's museum employees identified it as having come from an extinct wild pig. The latter information was kept under wraps until long after the Scopes trial. When Darrow declared in Dayton, "The scientific evidence is there for all to see," actually some of it was being held back. Indeed, as we shall see, a lot of it was being held back.

Creationist Duane Gish, a biochemist, joked about Nebraska Man: "I believe this is a case in which a scientist made a man out of a pig and the pig made a monkey out of the scientist." As one more chuckle, in 1972 the pig was found not to be extinct. Some were rooting around in Paraguay.

So many visitors poured into Dayton that the locals appeared to become a minority in their own town. Besides a horde of curiosity seekers, there were legal experts, science experts, Bible experts, newspaper reporters, radio reporters, and newsreel producers (with an airplane to carry out film every day). Most colorful, but unnerving to some visitors, were the mountaineers, all of the men carrying rifles as was their custom. (There might be game to shoot on the way into town. Such devotion to hunting produced the sharpshooter Sergeant York.) Some of the media people represented Britain, France, and Germany. Two of the world's greatest orators would battle on a matter of universal concern. More than a million words would be filed, and they would be largely misleading or untrue.

The visitors liked Dayton with its cheerful friendliness and Southern hospitality. Even the acerbic Mencken said, "The town, I confess, greatly surprised me. I expected to find a squalid Southern village . . . pigs rooting under the houses and the inhabitants full of hookworm and malaria. What I found was a country town full of charm and even beauty." That was at first. Mencken in his dispatches reverted to form, calling the inhabitants "yokels" and "morons." The accumulation of such insults made it so dangerous for him in Dayton that a city official urged him to leave. Mencken did so and missed the most exciting part of the trial.

The unjailed Scopes had a fine time driving VIPs around in a borrowed yellow roadster, reportedly once almost running into Mencken. The weather being extremely hot, the defendant went swimming with two of the prosecutors. Rapplyea fixed up an abandoned mansion, owned by the coal company, to house the expert witnesses. Some called it the "Defense Mansion," others the "Monkey House." Monkey dolls began appearing in store windows. The drugstore sold a "Monkey Fizz." One could buy a pin that said, "Your Old Man's a Monkey," and a chimpanzee cavorted on the courthouse lawn. The town was in a festive mood. As the trial approached religious signs went up.

Contrary to the story in *Inherit the Wind*, there were no ugly demonstrations against the teaching of evolution. Had there been dramatic demonstrations such as those in the play, there would have been no need for the fake shooting incident at the barbershop.

Why were the Daytonians so unconcerned about evolution? Once I happened to hear interviewed an elderly Dayton woman who remembered the trial. She said that evolution was not a big issue in the town because, even if it were taught, "Nobody would have believed it."

As mentioned earlier, the defendant never took the stand. Such a decision often is interpreted as a sign of guilt, but in Scopes's case it was a sign of innocence. If he had testified truthfully, it would have brought the trial to a sudden halt, and everybody wanted to have a trial. As for Scopes's excuse for not testifying, in New York he said that this was because he did not know much about evolution and he did not want that to reflect unfavorably on the case. This young American could be as slippery as Charles Darwin.

Doc Robinson gave evidence against the defendant, recalling that, according to Scopes, one could not teach biology without teaching evolution. The offending biology book, Robinson admitted, for years had been sold at his store. Darrow asked, "And you were a member of the school board?"

"Yes, sir," said the Hustling Druggist, drawing a laugh from the spectators.

Darrow said maybe somebody should advise Robinson that he was not bound to answer these questions.

"The law says 'teach,' not 'sell,'" interjected one of the prosecutors, and that drew a bigger laugh.

Of the seven boys who had testified to the grand jury, two were called upon at the trial, and one of them, being recalcitrant, had to be assured by the defendant that he would be doing him a favor. In his memoir Scopes said he told the boy "to testify to what he had been told to say." The reluctant witnesses answered enough questions to convince the jury that they had been taught about evolution. One boy's mother proudly informed the press that he had gotten out the biology book and "studied it up" before taking the stand.

In short, some high school boys were pressured into committing perjury because an evolutionist from New York, a few Dayton boosters, and the ACLU wanted to have a trial. Scopes felt that *his* "skirts were clear" because, after all, he *truthfully* had pleaded innocent. In *Center of the Storm* Scopes excused the boys' fictitious recollections by saying, "Yet I am sure they had not perjured themselves. Possibly they had read of the processes of evolution and thought I had taught it to them." He knew better. After the trial Scopes told a reporter friend that "the lawyers" had taught evolution to them.

Inherit the Wind tells us that no testimony from scientists was allowed, but put into the trial record by the defense was a statement from Dr. Winterton C. Curtis, a zoologist from the University of Missouri. He testified that the embryos of mammals had gill slits and that Bateson's speech in Toronto had been misunderstood. Paid $100 by the ACLU, Curtis made a pretty good hired gun. Reportedly, the professor returned the money after his financial situation improved.

The last day of the trial was the most exciting. Bryan agreed to testify in defense of the Bible if Darrow would testify in defense of evolution. Darrow's agreement to testify is not mentioned in the play or films, which largely rewrite Bryan's testimony so as to present him as a feckless idiot.

In real life as well as in the play, Darrow sarcastically asked whether Bryan believed in various Old Testament miracles. (Neither Darrow nor the dramatists were so devoted to science that they questioned New Testament miracles.) At great length Darrow hectored Bryan about how Jonah survived getting swallowed by a whale, how Elijah made the sun stand still, and so forth. Bryan saw no problem with God working miracles, and for that he ever since has been ridiculed in the media.

Reading the transcript it can be seen that Bryan, although handicapped by the miracles problem, handled himself pretty well. Asked where Cain got his wife, Bryan answered, "I leave it to the agnostics to hunt for her." When asked whether he had studied religions other than Christianity, Bryan, having visited China and India, explained why he thought Christianity more beneficial for humanity than Confucianism or Buddhism. Finding himself on the losing side of this exchange, Darrow objected to how Bryan was permitted to "regale the crowd with what some black man said to him when he was traveling."

A high point of both the play and the movie was the following dialog:

Bryan character: "A fine Biblical scholar, Bishop Ussher, has determined for us the exact date and hour of the Creation. It occurred in the year 4004 B.C."

Darrow character: "That's Bishop Ussher's opinion."

Bryan character: "It is not an opinion. It is literal fact . . . he determined that the Lord began the Creation on the twenty-third of October in the year 4004 B.C. at—uh, at 9 a.m."

Darrow character: "That Eastern Standard Time?" (Laughter from the spectators.)

The playwrights made a good joke, but it was fictitious. In reality, Bryan never vouched for Bishop Ussher's calculation, and neither did the bishop specify 9 a.m. In response to repeated questioning about the age of the earth, Bryan said that he did not know the age of the Earth and that the creation could have taken "millions of years." Unfortunately for the witness, the reference to millions of years angered his fundamentalist followers, who believed firmly that the Creation had required six twenty-four-hour days, not as Bryan thought, six "periods" of time.

Lead prosecutor was A. Thomas Stewart, the district attorney, and he kept trying to stop the interrogation as irrelevant to whether Scopes was guilty of teaching evolution, but Darrow and Bryan insisted on continuing. Judge Raulston indulged the two antagonists, and the proceedings become more of a debate than a trial. (Raulston was deferential toward Bryan. He also liked to get his picture taken, and perhaps that was a motive for stretching out the case.) Said Bryan at one point: "They [the defense attorneys] have not had much chance. They came here to try revealed religion. I am here to defend it, and they can ask me any questions they please."

According to Bryan, Darrow insulted the local people attending the trial, and Darrow, apparently annoyed that Bryan was getting more applause, responded, "You insult every man of science and learning in the world because he does not believe in your fool religion."

Perhaps Darrow lost his temper when making that response. However, it fitted what Sue Hicks perceived as Darrow's strategy, which was to be very provocative in order to get publicity and to goad the prosecutors into making an appealable error. The first part of the strategy was successful, probably beyond Darrow's wildest dreams. The second part was not successful.

After wrangling for approximately two hours, Darrow and Bryan got into a shouting match, and the proceedings abruptly were adjourned. The final, real life dialog:

> Bryan: "The only purpose Mr. Darrow has is to slur at the Bible, but I will answer his questions . . . . I want the world to know that this man, who does not believe in a God, is trying to use a court in Tennessee . . ."

> Darrow: "I object to that."

> Bryan: "to slur at it, and while it requires time, I am willing to take it."

Darrow: "I object to your statement. I am examining you
on your fool ideas that no intelligent Christian on earth
believes!"

Judge Raulston: "Court is adjourned until nine o'clock
tomorrow morning."

Inherit the Wind has the Bryan character linger at the witness
stand, babbling the names of the books of the Bible as his former
supporters desert him and crowd around the plainly triumphant
defense attorney. Next the Mrs. Bryan character cradles her
husband's head on her breast murmuring, "It's all right, baby. It's
all right." If any such babbling or cradling took place, it surely
would have been reported in the press, but I have found no trace
of it.

As for all the spectators deserting Bryan and gathering around
Darrow, the pro-evolution writer Watson Davis of Science Service
reported only that "there was a surge of young students, girls and
boys, to shake the hand of Darrow."

On the next day Judge Raulston barred further testimony by
Bryan, saying that it had no bearing on the guilt or innocence of
the defendant. Very true. Not only was the debate irrelevant, but
the jurors were excluded from hearing it.

Darrow said he had no witnesses to call, and the world's most
famous defense attorney, champion of the underdog, asked the judge
to instruct the jury to *find his client guilty.*

Prosecutor Stewart had no objection to that, and Judge Raulston bought the deal.

Bryan in that way lost his opportunity to put Darrow on the witness stand. He had plenty of ammunition. He could have attacked Darrow's vulnerabilities such as the Bateson speech and how the writings of Darwin and Nietzsche had led to the thrill killing in Chicago. There are differing accounts of why the judge ended the trial. One holds that Raulston made the decision for reasons known only to himself, another tells us that Stewart, over Bryan's protest, finally persuaded the judge to stick to the law. The *Inherit the Wind* movie blames political pressure from the state capital.

According to L. Sprague de Camp, who wrote a lengthy history of the event (*The Great Monkey Trial*), Dayton officials met secretly with the judge and urged him to stop the trial because feelings among the onlookers had risen to a fever pitch and somebody was likely to get hurt. Sue Hicks later said that the excitement of the crowd became "almost beyond the control of the small number of officers which we had at our disposal." Hicks blamed that on the defense's doing "everything they could to provoke the court and get on the front pages of the newspapers." Of course, it was extremely provocative for Darrow to use terms like "fool religion," and that surely was risky in view of crowd's temper and the sharpshooting mountaineers in town.

The Gang of Four was getting more excitement than it had bargained for.

In obtaining a quick end to the trial, Darrow was able to prevent his cross-examination, to keep the stemwinding orator Bryan from making a closing speech to the jury, to avoid a hung jury, and to obtain a guilty verdict that could be appealed on constitutional grounds. The folksy Chicagoan wanted to snap his colorful suspenders at higher levels of jurisprudence, hopefully even in Washington at the U. S. Supreme Court.

The jury returned a guilty verdict in nine minutes.

As his penalty Scopes could have been fined $500, but the judge required only $100. At this point *Inherit the Wind* pushes its slander to the limit. The Bryan character, outraged by the leniency of the penalty, demands a "more drastic punishment" in order to "make an example of this transgressor." In reality, Bryan always was friendly to Scopes, and he even offered to pay the fine.

While slanderous toward Bryan, the dramatization treats Mencken far better than he deserved. After the sentencing Scopes was free to go, but in the play he was required to pay $500 bail, and in the movie the Mencken character—not having fled for his life—announced his newspaper would take care of that.

Inherit the Wind shows the defeated Bryan, ignored by his former followers, collapse on the floor of the courtroom. It is implied that he died on the spot or shortly thereafter. Evidently not quite the crushed and broken man as has been claimed, Bryan after the trial actually went around giving speeches against Darwinism, and he assisted the planning for what became Dayton's Bryan College. Five days after the trial the Great Commoner passed away quietly while taking a nap.

"Well, we killed the son-of-a-bitch," was Darrow's reaction. Some people thought Bryan died from the severe July heat and the strain of the trial. Others blamed his diabetes combined with a habit of over-eating. (At a banquet in Dayton Bryan consumed half of Scopes's meal in addition to his own.) There was no autopsy.

Mencken wrote a vengeful obituary that his friend and biographer Sara Mayfield termed "savage." As Bryan's body lay in state, people filed by for two days. The deceased had been a colonel in the U. S. Army during the Spanish-American War, and so he was buried at Arlington Cemetery.

Scopes might have felt some sting from the conviction. While giving International News Service reporter William K. Hutchinson a ride in the yellow roadster, the driver unburdened himself of the fact that, although convicted, he actually was innocent because he never had taught evolution. Through the entire trial, said Scopes, he had been "scared" the boys might remember that they missed that lesson. Then the trial would have come to a halt, and he, Scopes, would have been "run out of town on a rail." Scopes pledged Hutchinson to keep quiet about all that until the case was disposed of.

At least one onlooker suspected that Scopes had not taught evolution. This was Mrs. Charles F. Potter, wife of the defense's Bible expert. She heard from Mrs. Rapplyea that Clarence Darrow was *coaching the prosecution's witnesses.* By that account Darrow suborned perjury in his most famous trial, indeed the most famous trial in American history. (Was that prosecutor in Chicago right to be suspicious about relations between the judge and the defense?)

Scopes told the whole truth to Mrs. Potter. He explained that his biggest job at the school really was to be the athletic coach, and he sometimes used the biology class to plan plays. "I reckon likely we never did get around to that evolution lesson," said he. Scopes did not skip evolution because of the Butler Act. By his own account in *Center of the Storm,* he was not concerned about that at all. However, after the trial in Dayton, nobody in Tennessee could forget that teaching evolution was against the law.

The *Inherit the Wind* play and films project a vivid stereotype in which the Bible is absurd and Darwin's theory of evolution is scientifically correct. Of course, the scientific objections to the story in Genesis do not prove the soundness of Darwin's theory. Nevertheless, the effect of the *Inherit the Wind* productions was to sell Darwinism and they still do.

In the aftermath of Dayton's big event, the town experienced a surge in religious fervor. The nastiness of Clarence Darrow and H. L. Mencken was diagnosed as the inevitable result of their agnosticism and evolutionism.

Charity was not foreign to the scientists. At the urging of "cherubic little" Watson Davis (de Camp's description) the defense's expert witnesses took up a collection among themselves and their colleagues and established a scholarship fund that Scopes could use for graduate school. (De Camp described the Science Service man to a T. I happened to meet Davis in the 1950s, and he, still cute as a button, brimmed with good will and good cheer.)

Scopes, a celebrity convict, was accepted by the University of Chicago, where he was expected to study biology. Instead he took up geology and became a petroleum engineer.

Scopes tried to be involved as little as possible in the ACLU's appeal. The opposite was true of Darrow, who clung to the case despite the ACLU's strenuous efforts to dump him. (Talk about arousing animosities. Dayton's district attorney branded Darrow "the greatest menace that present-day civilization has to deal with.") The ACLU-Darrow battle still was going on when the Tennessee Supreme Court rendered its decision. The court upheld the constitutionality of the Butler Act. But it reversed Scopes's conviction on a technicality concerning how the fine was decided. That meant there was no fine to pay and no conviction to appeal further.

The court urged the attorney general to dismiss this "bizarre case" on the grounds that the defendant, now in Chicago, was no longer in the state's employ. Besides, dismissal, said the court, would better serve the "peace and dignity of the state." The attorney general complied the next day, and there went the ACLU's Tennessee test case.

Besides selling Darwinism, the Scopes trial empowered the newly unforgettable Butler Act, which was not repealed until 1967. Meanwhile, the teachers coped with the law each in his or her own way. One way was to ignore human evolution. Another was for the class to go outside and sit on the lawn so that the instruction did not take place within the school building. Some teachers told their students to read that stuff at home. I have not seen this reported, but I would imagine that for a good while many teachers simply ignored the subject—as had the world renowned John T. Scopes.

Chapter Four

The Piltdown Men and Us

Piltdown man, a very early type of man (believed to belong to an earlier period than the Neanderthaloid type), whose existence is inferred from fragments of a skull discovered at Piltdown in 1912. American College Dictionary, 1953

How did we get from Homo erectus *to* Homo sapiens? *Book after book, conference after conference address themselves to this issue.* Harvard University physical anthropologist William Howells, *Getting Here*, 1993

Having never sought a new member, the gradually perishing X Club stopped meeting in 1893, but the youngest of the group, Sir John Lubbock (he inherited a title; so did Joseph Hooker), survived until 1913, the year of Alfred Wallace's death. This was long enough for Lubbock to be duped into putting his stamp of approval on the fossil forgery known as "Piltdown man," a "discovery" that was announced in 1912. This was an unfortunate conclusion to a very distinguished career. Lubbock did important work in archaeology and entomology, he coined the terms *paleolithic* and *neolithic,* and he saw to the preservation of Stonehenge, whose stones were being looted by builders. Lubbock's official biographer contended that

Lubbock's books, such as the extremely popular *Pre-Historic Times*, did more to make evolution credible than the more argumentative efforts of Thomas Huxley. If so, perhaps it helped to be an expert on tact.

Piltdown man, the "Science Fraud of the Century," enjoyed four decades of existence walking on the dark side of Darwinism with his wraith of a brother Piltdown II. On their heels came Piltdown III, a more modern looking human, equally fictitious. Until their exposure in 1953 the Piltdown men were ranked as one of the most important scientific discoveries of the twentieth century. The Piltdowners joyfully were claimed to provide the missing links between ape and man that scientists had been looking for.

The press bannered headlines such as "Darwin Theory Proved True," and anti-evolutionist members of the clergy were ridiculed. Three English scientists were knighted in connection with the Piltdown affair although the original discoverer, an amateur paleontologist, was not. The amateur was a local lawyer (solicitor) by the name of Charles Dawson.

With respect to Piltdown I, the forger (or forgers) had doctored up and put together the partial skull of a human being and the partial jaw of an orangutan. The bones were dyed to make them look older, and the teeth were filed to make the pattern of wear look human. Along with the apparent remains of Piltdown man, there were planted in a gravel pit prehistoric animal parts and what looked like ancient flint cutting tools.

The deception could have been discovered quickly with more thorough testing, even using the technology of 1912. The cranium was tested for mineralization but the jaw was not. Had the jaw been tested, it would have yielded organic matter indicating an age younger than that of the cranium. The forger had selected and treated the bones in such a way as to fit the pattern of human evolution—large cranium and small jaw—that scientists were expecting, and that fulfillment of prophecy enhanced the bones' credibility.

The Piltdown story is a striking example of how the intense desire to prove a theory can produce a mass delusion among scientists. This might not have happened if the American paleontologist Othniel Marsh still had been alive. The Yale professor was an avid detector of hoaxes. One was the ten-foot Cardiff Giant, supposedly unearthed near Syracuse, New York. People bought tickets to see evidence for the book of Genesis's report of "giants in the earth." Marsh noticed that the allegedly petrified man had been carved from gypsum, a material that would have dissolved when the earth was wet.

Also assisting Piltdown's acceptance was ethnic pride. When discovered, the local hominid was believed to be older than the Neanderthal man found on the continent of Europe, and the Piltdown man, with his high forehead, looked more human and less apelike than the European hominid. The Neanderthal, therefore, was alleged to be a degenerate offshoot of the Piltdown species, and many a Briton could see the logic of that.

One of the knighted paleontologists, Arthur Smith Woodward, stoked ethnic pride by writing a book titled *The Earliest Englishman*. Helping to justify that appellation was the discovery of a flattish elephant bone shaped like a cricket bat. Some might have thought that comical, but another of the knighted scientists, the anatomist Arthur Keith, judged this artifact "the most amazing of all the Piltdown revelations," and in a foreword to *The Earliest the Englishman*, Keith declared, "No theory of human evolution can be regarded as satisfactory unless the revelations of Piltdown are taken into account."

In the panicky aftermath of the forgery's discovery, fingers pointed in all directions. No fewer than twenty different men, including Woodward and Keith, were accused of having been responsible for the hoax.

One of the alleged forgers was Sir Arthur Conan Doyle, the writer of the Sherlock Holmes mysteries, who resided only about eight miles from Piltdown. Doyle was a medical doctor, and his fiction betrayed some knowledge of paleontology (for that matter, his Sherlock Holmes character betrayed some knowledge of whatever came along). Piltdown man was announced to the public about when Doyle's novel *The Lost World* appeared, and the astonishing discovery boosted sales of Doyle's story about apemen and dinosaurs surviving together on an isolated plateau in South America. Doyle's apemen, however, did not look like the Piltdown species. They were more apish and smaller brained.

Critics of the Doyle accusation could not imagine a famous author repeatedly sneaking in and out of the Piltdown gravel pit. As for Doyle's expertise in paleontology, it was not much. Charles Dawson knew far more. He was acquainted with the novelist and gave him pointers on the subject.

Also accused was Father Teilhard de Chardin, a young Jesuit priest and budding paleontologist. Teilhard in the summer of 1913 was studying at a seminary not far from Piltdown. He and another student volunteered to help excavate the newly famous gravel pit, and Teilhard found a human-looking tooth. Stephen Gould made a circumstantial case against the priest in his 1980 book *The Panda's Thumb* and again in his 1983 book *Hen's Teeth and Horse's Toes*. As Gould pointed out, Teilhard had been acquainted with Dawson, a fellow fossil collector, for a few years prior to the Piltdown discoveries.

John E. Walsh, author of *Unraveling Piltdown*, looked into Gould's case against Teilhard and was unpersuaded. Ronald Millar, author of *The Piltdown Men,* thought it suspect that Teilhard found in the gravel pit the radioactive tooth of an extinct elephant. Such radioactive teeth, said Millar, came only from Tunisia, which Teilhard had visited. Thomas M. King, S.J., wrote a point by point rebuttal to Gould's charges in his book *Teilhard and the Unity of Knowledge*. According to Father King, Dawson had a stepson with the British army in Africa and received from him many objects of interest.

Also accused was Dawson. He discovered the first fossil, which was part of a skull, and most of the rest. Dawson reported the initial discovery to his friend Arthur Smith Woodward, who was Keeper of Geology at the British Museum of Natural History. Dawson for years had been bringing artifacts and bones to the attention of Woodward, and the latter regarded his work with respect. Woodward wrote of Dawson upon the lawyer's death:

> He had a restless mind, ever alert to note anything unusual; and he was never satisfied until he had exhausted all means to solve and understand any problem, which presented itself. He was a delightful colleague in scientific research, always cheerful, hopeful, and overflowing with enthusiasm.

Woodward himself was not one to overflow with enthusiasm. He was very reserved and wore a Vandyke beard. But the two men got along well.

According to Dawson, around the year 1908 (the circumstances of Dawson's discoveries tended to be vague) he noticed some odd flints uncovered by laborers who were repairing a road with material from a gravel pit, and he asked the men to preserve anything they might find. On a later visit one of the laborers reportedly gave him an unusually thick piece of skull bone. Next, in the autumn of 1911, Dawson himself found in the area a larger piece of skull, also rather thick. In 1912 Dawson and Woodward together began to excavate the gravel pit with the help of an elderly laborer named Venus Hargreaves (male, like Sue Hicks), who became another of the men accused of complicity in the forgery.

Somebody took a photograph of the three men at the pit, and dominating the foreground was a large white goose named Chipper. An aggressive bird, Chipper was complicit in that he helped to keep curiosity seekers away.

Once as Woodward was looking on, Dawson whacked the gravel with a geologist's hammer and up flew the piece of a jawbone. This was not an accepted excavating technique, but it got an important, though fake, result.

In December of 1912 Woodward and Dawson presented their findings to the Geological Society in London. Thanks to press reports of what the two had discovered, this meeting of the society was the best attended and most exciting held up to that time. As the most human-looking of our evolutionary predecessors, Piltdown man constituted by far the most convincing evidence of humankind's apelike ancestry.

Woodward and Dawson displayed nine pieces of cranium and the partial jawbone, which retained two molar teeth. The thick-walled cranium looked human with some simian features while the jaw appeared to be simian with human features (the teeth). The upper surfaces of the teeth apparently had become flattened by chewing in the human manner, which includes jaw motion from side to side. Apes chew only up and down.

What Woodward named *Eoanthropus dawsoni* (Dawson's Dawn Man) was declared to be a new genus in the human line of descent. Going by associated animal fossils and geological stratigraphy, Dawson estimated the biped's age as 500,000 years. Somebody suggested a million years, but Dawson stuck to his conservative figure.

Piltdown was a paradox of man and ape, and this is just what scientists had been expecting. Many of those present had heard Thomas Huxley lecture about the apish ancestors of humankind, and Darwin in his book *The Descent of Man* had described our evolutionary forebears as having had "great canine teeth which served them as formidable weapons." The brain of the Piltdown man, having advanced more rapidly than the face and jaw, agreed with contemporary ideas. The anthropologist Grafton Eliot Smith, who would become one of the Piltdown knights, declared that the reconstructed cranium had encased the most primitive and apelike human brain yet found. But he said, the skull showed signs of incipient expansion (faint praise for the human decedent).

There were objections. Some doubted that the jaw and cranium belonged to the same individual. Missing from the jaw was the condyle (swivel knob) needed to hold the jaw to the cranium; and would it not be odd for an apelike jaw to possess a human-type condyle? Woodward argued that the fossils had been found close to each other, that they were similar in color and mineralization, and that the teeth indubitably were human-like. Woodward agreed that if a canine tooth were found, it would resemble the large tooth of a chimpanzee but show a different kind of wear.

It took eight months to find the tooth. Teilhard spotted it near where the jawbone had been discovered, a likely spot which, strangely, did not seem to have been looked at carefully. The tooth was large, and the wear was humanlike. This strengthened considerably the case for the jawbone and cranium having been parts of the same individual.

Scientists brushed aside the criticism of a dentist who said that the tooth was immature and therefore not consistent with its large amount of wear.

The elephant bone, whether a club or a cricket bat, turned up in 1915. The smaller end had been rounded with a cutting tool, apparently to make it easier to handle.

At the time of the bone's discovery Teilhard was in France. What was called the Great War had begun and he was serving as a stretcher bearer. After his younger brother was lost in action, the priest, refusing a commission, volunteered as an enlisted man, and he served until the end of the war, receiving several decorations. Afterward Teilhard became prominent as a paleontologist and philosopher. In his philosophy the Jesuit attempted to reconcile religion and evolution, and he attracted a large following.

One of the Piltdown skeptics was an American, Gerrit Miller, who was curator of mammals at the United States National Museum. Miller decided the jawbone was that of a previously unknown kind of chimpanzee. Woodward dismissed Miller's opinion as "the latest

ROT from the U.S.A." One of Woodward's colleagues published a lengthy rebuttal of Miller, alleging the American had set out "to confirm a preconceived theory."

Piltdown III, a more modern human, was discovered before Piltdown II (which was similar to Piltdown I). In 1913, Dawson reported to Woodward that he had found pieces of a more modern looking skull at a place called Barcombe Mills, and he suggested that the new man, having a skull of normal thickness, could have been an evolutionary descendant of Piltdown I. There is no record of a reply from Woodward. But he entered the discovery in the formal register of the Natural History Museum. The antique-looking fragments came to scientists' attention in 1949 and were accepted by some as an evolutionary descendant of Piltdown I. Others thought Piltdown III was really just a Neanderthal. (How could there be such a wide range of opinion?)

In 1915, the most conclusive discovery was reported, and this is what became known as Piltdown II. Dawson informed Woodward that two miles from the famous gravel pit he had found a molar tooth and a piece of cranium similar to what had been in the gravel pit. This news pretty well clinched the Dawson-Woodward case for their earliest Englishman.

Dawson fell ill in 1915 and died the following year. Some thought it unfair that the amateur paleontologist was neither knighted nor elected to the Royal Society, an honor that he particularly had coveted. Life treated Mrs. Dawson even more unfairly. She was a wealthy widow when she married Dawson but was left in straitened circumstances. Dawson had neglected his law practice in favor of his scientific hobbies. Woodward obtained for Mrs. Dawson a government pension, presumably in recompense for her husband's efforts on behalf of science. (It will be recalled that Alfred Wallace received a government pension. However, it was cut off when he died, and that left *his* widow in straitened circumstances.)

There still was skepticism in America, and one of the most unconvinced was Henry Fairfield Osborn. After the Great War ended, Osborn went to Britain, conferred with Woodward, and examined the fossils. This resulted in a dramatic change of opinion that has been compared to a religious epiphany, and the museum president's sponsorship of Piltdown man carried a lot of weight among paleontologists. (The museum, of course, was a source of funding.) Concerning the Piltdown II specimens, Osborn marveled that "they were exactly those which we should have selected to confirm the comparison with the original type." They were exactly those which a forger would have selected too.

The Piltdown men definitely had arrived.

After Sir Arthur Smith Woodward retired from the museum, he moved to the Piltdown area, where he spent the next twelve years looking for more fossils. He did not find any. In 1938, Woodward, his health declining, gave up on fossil hunting and arranged for a large memorial stone to be placed where the first fossils had been found. Sir Arthur Keith unveiled the stone and gave a speech lauding

Dawson for his discovery, which Keith compared to other important advances in science that had encountered stiff opposition before they were accepted.

But Piltdown's reign would be limited. In 1948, a new way of determining the age of fossils, a test for fluorine absorption, showed that the cranium and jaw were at most only about 50,000 years old. That was a time when *modern* looking men were walking around. This meant that the Earliest Englishman, far from leading the evolutionary parade, was a laggard, perhaps even (horrors!) a *degenerate*. Besides that, the accumulation of other hominid fossils continued to look very apelike with their low foreheads. Instead of a missing link, Piltdown man now looked more like an orphan.

In 1953, there was held a scientific conference in London on the subject of early man in Africa, and one of the attendants, a physical anthropologist by the name of Joseph Weiner, took the opportunity to visit the British Museum of Natural History and view the famous fossils from Piltdown. Ever since the fluorine test, Weiner had been puzzled by the Piltdown discoveries. At the conference banquet, Weiner found himself sitting next to the paleontologist, Kenneth Oakley, who had developed the fluorine test and examined the fossils. Weiner asked Oakley why the Piltdown II site never had been fully excavated. The explanation was simple: nobody knew where it was.

"The fact is," said Oakley, "all we know about Site II is on a postcard [Dawson] sent in July 1915 to Woodward, and an earlier letter from that year, from neither of which can one identify the position of Piltdown II."

Weiner lay sleepless that night. Piltdown II had convinced the world of Piltdown I's authenticity. But nobody knew the circumstances of the 1915 discovery or where it took place. Why had he never heard this before? Had even Woodward, now deceased, known where the site was?

Early the next morning Weiner went to his laboratory at Oxford University and examined carefully plaster casts of the Piltdown fossils. It did not take long for him to suspect forgery. The teeth did not look right. The edges of the molars were sharp as if filed down instead of worn down by chewing. The flattened surfaces were canted in opposition directions, and the degree of wear on the two molars was exactly the same although the inner molar normally wears down more quickly than the outer one.

Looking at the report of Oakley's fluorine test of a tooth, Weiner learned that just beneath the darkly stained surface the dentine was pure white not stained as it ought to have been after being buried in the ground for thousands of years. Weiner decided to stain and file a fake fossil tooth of his own, and he found that this was surprisingly easy.

Reading Dawson and Woodward's report from 1912, the Oxford investigator noticed that the jaw had not been actually tested for mineralization. This was an amazing lapse in proper procedure, especially considering there originally was much skepticism concerning whether the jaw and cranium really belonged to the same individual.

Weiner telephoned Oakley about his suspicions and asked him to put the actual Piltdown teeth under a microscope and look for marks of filing. Only an hour passed before Oakley called back to say that yes, the teeth had been filed.

Discovering the forgery was quick and easy, once somebody took a careful look at it. Before announcing publicly the shocking news, much additional testing was done, and in a few months it was proved that *all* of the Piltdown fossils, including the various animal remains, had been faked in some way. As for the thickness of the cranium, this was determined to be a pathological condition perhaps caused by anemia. Rickets also was suggested.

Another surprise: Piltdown II never actually existed. Those fragments had come from the Piltdown I skull planted in the gravel pit. No wonder Osborn said that between the first and second group of fossils "there is not a shadow of difference."

The controversial jaw, it eventually was decided, was that of an orangutan. The new process of carbon 14 dating established its age as 500 years, plus or minus 100 years. The cranium was 620 years old, plus or minus 100 years. Piltdown Man was the combination of a medieval Englishman and an Asian ape.

At the age of eighty-six Sir Arthur Keith was still alive and residing in the Kent village of Downe, where Charles Darwin lived most of his adult life. Weiner and Oakley drove there in order to inform Keith of what had taken place and, as they hoped, to soften the blow. Keith, however, already had gotten the news from the London *Times* and was not looking well. The anatomist recalled that he initially had been very skeptical of the discovery, but additional fossils kept turning up in support of it.

Having exposed Piltdown man as an elaborate but clumsy fake, Weiner went a step further and tried to find out who was the perpetrator. First of all, he wondered, what do we know of Charles Dawson?

Woodward had respected the industrious amateur, and Keith said he had a reputation for great conscientiousness and accuracy. While still a schoolboy Dawson was presenting reptile fossils to the British Museum, and it was quite a distinction that at the age of twenty-one he became a fellow of the Geological Society. Dawson found an unknown mammal, which was named *Plagiaulax dawsoni*, and the fossil of an unknown dinosaur, which was named *Iguanadon dawsoni*. For more than thirty years he was an honorary collector for the British Museum. Needless to say, the "Wizard of Sussex," as he came to be known, was a member of the Sussex Archaeological Society. As an authority on old ironwork, the wizard submitted papers to the Society of Antiquaries, too, and he became a fellow of that organization. As an historian Dawson wrote a two volume *History of Hastings Castle*.

Would a highly respected Renaissance man want to risk his reputation by committing fraud?

Before the Piltdown forgery was announced to the public, Weiner drove down to Lewes, Dawson's hometown. There was located the Barbican Museum, home of the Sussex Archaeological Society, and Weiner expected the place to be bursting with Piltdown lore. To his surprise, very little was on display. There were only a cast of the skull, a small artist's version of Piltdown man presented by a Dr. Spokes, and a few flints from Piltdown presented by one Harry Morris. A card indicated that the cast had been there since it was donated in 1928. Apparently no cast was displayed until long after the world's most famous hominid had been accepted by the world's leading scientists. And, as another surprise, Dawson, despite his extraordinary discoveries, never became a member of the society's governing council.

Weiner expected to find extensive coverage of the Piltdown discoveries in the annual *Sussex Archaeological Collections*. But there was said nothing at all until 1925 when the *Collections* mentioned that Sir Arthur Smith Woodward had given a talk on the Piltdown man. This ought to have been a stellar occasion, but the text of Woodward's remarks was not printed. In 1916, the death of the society's world famous member went unreported in the *Collections* although Dawson was eulogized in the commercial press.

Evidently to the Sussex Archaeological Society, Charles Dawson was a pariah. Investigations by Weiner and others uncovered many reasons for that.

First of all, it must be noted that the society ever since 1904 had held a grudge against the famous archaeologist-paleontologist-historian-antiquarian. In the view of members, Dawson had purchased underhandedly their previous headquarters for his own use as a residence. The society had been meeting since 1885 in that building, Castle Lodge, which formerly had been a part of Lewes Castle, and members of the society were under the impression that if the owner of the building ever wanted to sell, they would be allowed to buy. To their consternation, in 1904 they suddenly received an eviction notice. The agent explained that, until the final stages of purchase, he thought Dawson, a solicitor and a member of the Archaeological Society, was acting on behalf of the society.

Dawson subsequently joined the Hastings Natural History Society and tried to avoid members of the Sussex Archaeological Society. That proved difficult because the society moved into new quarters near Castle Lodge.

So far as the Piltdown fossils were concerned, at least some of the society's members had suspected that Dawson was "salting the mine." One was Harry Morris, the flint collector. He believed the Piltdown discoveries to be fraudulent because the flints said to have been found there by Dawson were not typical of that area, and to him they did not look so old as Dawson claimed. Weiner learned of several other local skeptics. Reportedly, two of them entered Dawson's law office without knocking and found the solicitor staining bones. In explanation, Dawson said that he was trying to find out how bones were stained in nature.

There were various other complaints about the archaeological society's world famous member. Locally, Dawson was known as a man of so many doubtful distinctions that, as I would guess, the Sussex Archaeological Society must have held a special meeting for a good laugh when that cricket bat turned up. City folk ought to be more careful when world famous events take place in rural communities like Piltdown, England, and Dayton, Tennessee. The species *Homo sapiens* comprises country slickers as well as city slickers. I am reminded of an old television series, *The Ghost and Mrs. Muir.* At the fictional village of Schooner Bay, the leading citizens habitually gathered in the back of an antique store to discuss affairs of the day, while the antique dealer, an essential part of the tourist industry, busily distressed his wares. This was an harmonious arrangement except when the dealer used buckshot for the distressing.

Here are some of Dawson's other accomplishments:

Dawson's two volume *History of Hastings Castle* was viewed as a useful work, but the most informative parts were plagiarized from an 1824 study. Errors were frequent, and references often were given in unintelligible form. It was characteristic of Dawson that his sources and documentation were vague.

Dawson reported finding an extra vertebra in the skeleton of an Eskimo, and he asserted that the skeleton was a new race of men. This he never wrote up as a scientific paper, but he gave the story to the popular press.

Entrusted with excavating the Lavant Caves, Dawson did a sloppy job, damaged the site, and never produced the scholarly paper he had promised.

Dawson spotted a sea serpent in the English Channel.

He wrote an article, largely plagiarized, on old iron objects.

Dawson bought a three-inch iron statuette and claimed it to be a Roman work of art in cast iron. This became a big issue since cast iron was unknown in Britain at the alleged time of manufacture. Dawson told little of the object's origin. He said he bought it from a workman who had found it in an old Roman cinder heap, whereabouts unknown to Dawson. Many years after the original announcement, Dawson came up with a fuller story, which he wrote up for the Sussex Archaeological Society. He said that he again had encountered the workman who discovered the artwork and that he had obtained a written account of the discovery. But the written account never was produced.

Dawson discovered an allegedly Roman horseshoe unique in having holes for fastening by nails. The Roman custom was either to leave a horse unshod or to tie to the hoof a slipper of metal or leather.

Dawson found a clockface from the Middle Ages with human figures painted on it. Somebody noticed that the clothing was anachronistic.

The solicitor discovered a small mace that supposedly had been carried centuries before by a Hastings "water-bailiff." In time the mace was reclassified as a nineteenth century artifact.

Concerning the authenticity of *Plagiaulax dawsoni* and *Iguanodon dawsoni,* in 2006 I asked the British Museum of Natural History whether these finds still were considered authentic. The museum said that *Iguanodon dawsoni* is considered authentic, but that *Plagiaulax dawsoni* is "probably *nomen dubium.*"

Did Dawson have an accomplice? Several have been suggested, from Venus Hargreaves to Sir Arthur Conan Doyle. Dawson, however, seems not to have had an accomplice prior to the Piltdown caper, and the mine-salting halted with his death. For a helper, the Wizard of Sussex either relied upon some fictitious "workman" who could not be found or some very respectable person, like Sir Arthur Smith Woodward, who could be duped.

At the time Weiner became suspicious of Piltdown man, scientists believed that there could have been only one line of human evolutionary descent as the less competitive hominids became extinct by natural selection. That was Darwin's theory. And so as more apelike hominids kept turning up, the high-browed earliest Englishman, off by himself, became less and less credible.

Ironically, in recent decades the added accumulation of hominid fossils has tended to show that there probably *was* more than one line of descent—if in fact there were any *line* of descent at all. We now hear about "mosaic evolution." This sounds to me like Huxley's saltationism, which Darwin equated to "miracle."

Nevertheless, we still see published sketches showing the old familiar parade of bipedal figures getting taller, straighter, and better looking. One ends with a handsome, wavy-haired Caucasian. Viewers assume that this parade represents a Darwinian line of evolution, each biped evolving into the next, but that is an outdated notion based on theory. The parade really shows only an unexplained trend. Gould complained that foreign publishers stuck this sort of misleading illustration into four of his books without permission.

In 1996, paleontologists established the surprising fact that *Homo erectus, Homo neanderthalensis,* and *Homo sapiens* all coexisted just 30,000 years ago. (Might that account for the Genesis report of "giants in the earth"?) Added to the mosaic in 2003 were the remains of miniature humans, *Homo floresiensis*, which date back only 13,000 years. This hobbit-like person, with a head the size of a grapefruit, dwelled on the Indonesian island of Flores among pygmy elephants and the poisonous Komodo dragons.

DNA and skeletal evidence have established pretty well that the Neanderthal was not our ancestor. This leaves *Homo erectus* as a candidate, but he was very different from modern humans, and in Darwinian theory there have to be credible intermediate forms. According to the Harvard University anthropologist William Howells, with *Homo erectus* we confront the "greatest current problem in the human story." In his book *Getting Here,* the professor asked, "How did we get from *Homo erectus* to *Homo sapiens*? Book after book, conference after conference, address themselves to this issue." That was in 1993. In 2007, new evidence showed that *Homo erectus* was even less humanlike than had been thought.

In Kenya paleontologists found for the first time the skull of a female *Homo erectus*, and, significantly, it was much smaller than expected. In modern humans the male on average is 15 percent larger than the female, but the difference among gorilla male and females is much greater—as in the case of *Homo erectus*.

As another new development, the *origin* of *Homo erectus* has come into question. Fossils used to indicate that he evolved at about the same time that an earlier hominid, *Homo habilis*, became extinct. That made sense in Darwinian theory. However, in 2007 it was announced that more recent fossils of *Homo habilis* had been found, and that the two species lived at the same time and even near each other for almost half a million years.

The lead author of a report in *Nature* (August 9, 2007) was Fred Spoor, who told the *New York Times* that the new findings contradicted the notion of human evolution "as one strong, single line of evolution from early to us." The revolutionary fossils were found years earlier, but wrote John Noble Wilford in the *Times*, "The implications were considered so profound that little was said about these dates [ages of the fossils], pending more conclusive evidence." In other words, scientists were embarrassed by facts that they preferred not to reveal as happened in the case of Henry Fairfield Osborn's Nebraska Man.

Scientifically, the origin of the modern human looks more mysterious now than it did in the time of Charles Darwin.

Chapter Five

Wars and Genocides

German Darwinism was shaped by Ernst Haeckel, who combined it with anti-clericalism, militaristic patriotism, and visions of German racial purity. He encouraged the destruction of the established church in Germany, with its sermons about 'the meek shall inherit the earth' and compassion for unfortunates. Such a 'superstitious' doctrine would lead to "racial suicide." Richard Milner, *The Encyclopedia of Evolution,* 1990

In California overlooking Evolution Lake there stand three lofty mountains named after Charles Darwin, Thomas Huxley, and Ernst Haeckel. Few Americans today ever heard of Ernst Haeckel, but as one measure of his importance, Haeckel's mountain at 13,481 feet is taller than Huxley's, though shorter than Darwin's. Other mountains were named for Wallace, Lamarck, Mendel, Spencer, and even John Fiske. The last achieved distinction as a proponent of theistic evolution, which acknowledges a divine role. Fiske also believed that as the mind of man evolves, it becomes an indication of the mind of God.

Like Huxley, Ernst Haeckel was a firm believer in secular evolution but not in Darwin's theory. He rejected the idea that variations came about merely by chance. To his Prussian mind, evolution proceeded in an orderly manner by a set of fixed laws (an impression anyone might get by studying embryology). To quote Walter J. Bock in *Science* of May 9, 1969, "No indications of chance or random mechanisms were apparent to [Haeckel], nor was [he] aware of any processes that would introduce chance into evolutionary mechanisms."

Despite his differences with Darwin, in Europe Haeckel was the Englishman's chief ally and most enthusiastic admirer. He named one of his daughters after Emma Darwin, who, nonetheless, dreaded his visits. Haeckel knew little English, "roaring" described his conversational style, he disliked the Christian church, and as if theology were a branch of taxonomy, he classified Emma's beloved God as a "gaseous vertebrate." (Let us hope Emma did not find out about that.) Haeckel was an atheist or, to be more precise, a pantheist. As a nature worshipper, he found his god visible all around him.

Emma's husband liked Haeckel, finding him "pleasant, cordial, and frank." But Charles, too, was uncomfortable with the German's anti-clericalism. One of Darwin's best friends was the local rector, John Brodie Innes. Unlike Haeckel and Huxley, Darwin never sought conflict with the church, and nobody could maneuver him into it although some tried.

Haeckel invented the words *ecology, phylum, phylogeny,* and other biological terms.

While Huxley and his X Club were Darwinizing the English-speaking world, Haeckel took up the conversion of Europe, especially the German-speaking part of it. The German Darwin proudly mailed to the English Darwin lists of significant converts to evolution. But his influence was not confined to Europe. The Extraordinary Professor of Zoology at Jena University became the world's best selling writer about evolution and its implications. His books far outsold Darwin's. The world's sophisticates needed to have one or two lying around.

Important but not a big seller was Haeckel's two-volume *General Morphology*, which reorganized all of biology along evolutionary lines. With the help of a German dictionary Darwin struggled to read the ponderous work, grumbling that the Germans "could write more simply if they chose." One of Haeckel's biographers, fellow German Wilhelm Bolsche, rated the book a "model of clearness."

Many of Haeckel's countrymen took to *Darwinismus* like ducks to water. Large numbers of students attended Haeckel's lectures, and the charismatic professor also attracted to the university financial donations. Because of Haeckel's influence, Prussia awarded Darwin a knighthood, an honor he never received in England. From Bonn University in the Rhineland, Darwin received an honorary doctorate. He also was very pleased to receive an album bearing the signed photographs of one hundred and fifty German men of science. Dr. Darwin wrote to William Preyer, a physiologist at Bonn, that the enthusiastic German response was the "chief ground for hoping that our views will ultimately prevail."

The French reaction to Darwin was quite different—actually, nil. There he came up against a "conspiracy of silence," as Huxley put it. The French stuck to the ideas of Lamarck and to those of another French biologist, the Baron Georges Cuvier. Among other accomplishments, Cuvier established the amazing fact of extinction. Previously people assumed that the bizarre animals discovered by the paleontologists had simply gone somewhere else.

Before taking up biology, Haeckel was an art student, and he perhaps was the first to publish a genealogical tree of life. The gnarled branches featured species in the place of twigs, and from the very top, just above the apes, the *Menschen* peered out. Haeckel combined his artistic ability with a knowledge of marine biology to produce beautiful pictures of marine life, and that helped to inspire the Art Nouveau movement (which devolved into the simpler Art Deco style). When in 2005 a big Art Nouveau exhibit came to the National Gallery in Washington, D.C., it showed some drawings by Haeckel as examples of his influence.

Some Art Nouveau pictures spiced their colorful, swirling beauty with the inclusion of a demonic creature. Haeckel, too, had a demon within him. Carried away with his enthusiasm for *Darwinismus*, he faked evidence for it. He drew ape skeletons so as to make them look more human, but as a forger his greatest success was achieved with embryo drawings. It was Haeckel's contention that an organism's evolution was repeated in the womb, and his drawings of human embryos showed gill slits. He was not the first to portray such slits, and perhaps he even believed in them. In *The Riddle of the Universe*, a book published in approximately twenty-five languages, Haeckel said the "openings" disappeared "after a time." In addition to the gill slit misinformation, Haeckel drew the beginning stage human embryo along with the beginning stage embryos of various animals so that they all looked the same, even though in reality they differed. It is for this that he has been most censured, with the gill slits getting overlooked.

When at Jena Haeckel was accused of fraud, he tried to justify his work by saying that all biologists schematized drawings to make them more understandable, but his colleagues decided that the Extraordinary Professor had gone too far. The defendant was accused of plagiarism in addition to forgery. The trial was not just a local affair but a worldwide scandal reported in the *New York Times* of January 3, 1909.

Notwithstanding all the negative publicity, Haeckel's embryo drawings were adopted by textbooks and reference books, and their long continuance in print has been truly astonishing. Gould was fooled by them, probably in the 1950s when he studied biology in a Queens, New York, high school. In 1989, embryo drawings based on Haeckel's work (and so credited) appeared in the advanced textbook *Molecular Biology of the Cell*, although one of that book's coauthors was Bruce Alberts, who in 1993 became president of the National Academy of Sciences. Recently, as mentioned earlier, there appeared references to gill slits in a major college level textbook of 2005 and in a 2007 edition of the *National Geographic* magazine. Even in a book on cosmology I have found a discussion of human gill slits. Surely the audacious Prussian never expected that degree of success.

Haeckel's forgeries of embryo similarities have been exposed over and over again. After the trial at Jena there was published in 1915 the illustrated book *Haeckel's Frauds and Forgeries* by J. Assmuth and Ernest R. Hull. Creationist scholars routinely have complained about Haeckel's influence on public school textbooks. In 1997 a British embryologist, Michael Richardson, wrote an expose for the scientific journal *Anatomy and Embryology* published by a German company. That led to an article in the American journal *Science* titled, "Haeckel's Embryos: Fraud Rediscovered." The author, Elizabeth Pennisi, said, "Generations of biology students may have been misled by a famous set of drawings published 123 years ago by the German biologist Ernst Haeckel."

May have been misled? A more trenchant view was expressed by Gould, who wrote an expose for the *Natural History* magazine of March, 2000. He declared that Haeckel's embryo drawings were "the single most familiar illustration in the history of biology." While not criticizing the nineteenth century textbooks, he said, "We do, I think, have a right to be both astonished and ashamed by the century of mindless recycling that has led to the persistence of these drawings." Such illustrations had appeared in at least fifty recent biology texts, according to a letter Gould received from Michael Richardson. The letter was dated August 16, 1999.

"Mindless recycling" is an interesting term when one considers that biology textbooks are reviewed before publication by large numbers of scientists. In one book I counted the names of more than a hundred reviewers. Do scientists endorse textbooks the way athletes endorse shoes and breakfast cereals?

Curiously, Richardson in his letter to Gould criticized the historians for not telling the scientists about Haeckel's fakery. Who knew that historians were expected to help out with the self-correcting nature of science? Perhaps the historians ought to bestir themselves and tell the biologists about Louis Pasteur and his experiments concerning spontaneous generation.

In the *Natural History* article of March, 2000, Gould again told of human gill slits.

In addition to the matter of Haeckel's drawings, some scientists for decades had been conducting a crusade against Haeckel's contention that evolution repeated itself in the womb. Haeckel called this the "biogenetic law," and he expressed it as "ontogeny recapitulates phylogeny," a jawbreaker that still has not disappeared entirely from American textbooks. The biogenetic law began to be discredited as early as 1901. Nevertheless, George Bernard Shaw believed in it. So did Sigmund Freud, who thought the biogenetic law helped in understanding the human psyche. While writing this chapter I found abandoned in a library copying machine a paper on recapitulation theory and human infancy, which had been published in 1914 by Columbia University. I had to wonder whether the scholar who left the paper behind was still another victim of Ernst Haeckel.

Among research scientists the biogenetic law was pretty well discredited in 1922 by the embryologist Walter Garstang and again in 1930 by the staunchly Darwinian embryologist Sir Gavin de Beer. De Beer said that too many stages of evolution were missing; other stages appeared out of order or originated in a different embryonic organ. In 1940 de Beer remarked on the fallacious theory's amazing longevity, and he did so again in 1958, complaining. of how this "puerile notion" had thwarted and delayed the proper study of the embryo. "Seldom," wrote the exasperated embryologist, "has an assertion like that of Haeckel's 'theory of recapitulation,' facile, tidy, and plausible, widely accepted without critical examination, done so much harm to science."

How do we account for the phenomenal success of Ernst Haeckel as a forger of scientific documents and the propagator of a false theory? Perhaps we ought to call it the "Piltdown effect." Sometimes scientists are so eager to prove a theory that by some Orwellian mechanism they manage to delude themselves even in the face of unimpeachable contrary evidence.

The biogenetic law is identified with Haeckel, but Darwin agreed with it, according to his autobiography. He not only agreed with it but felt cheated concerning the priority. In Darwin's opinion his earlier comments on embryo similarities in *The Origin of Species* ought to have received more attention. But the wily Englishman admitted that he had failed on that subject to impress his readers and concluded (perhaps with an eye to Wallace *et alia*) that "he who succeeds in doing so deserves, in my opinion, all the credit."

The harm Haeckel did was not only scientific. With Teutonic thoroughness he decided to apply the laws of nature to politics and social behavior. In one way this encouraged democratic values since our allegedly apish ancestry tended to deflate the pretensions of the aristocracy. However, according to Haeckel, politics was "applied biology" and so it was the natural duty of the strong to overcome the weak. The strong, of course, was Haeckel's own Aryan race, which already was dominating most of the planet. A supporter of eugenics, Haeckel demanded that his race be kept pure.

Anti-Semitism was part of the Prussian's philosophy. He argued it "credible" that Jesus Christ was not purely Semitic but partly Aryan because of the Nazarene's "high and noble personality." For Jesus's Aryan ethnicity Haeckel provided documentary evidence: apocryphal gospels reported that Jesus's real father was a Roman officer. Some Roman officers were ethnically Greek, and Haeckel was pleased that Jesus's alleged father bore the Greek name of Pandera (noting the similarity to Pandora of the Greek myth about Pandora's box). The Hellenes, said Haeckel, were the highest form of Aryan.

In 1906, Haeckel founded in Jena the Monist League, an organization consisting largely of university professors, which favored the purity and dominance of the Aryan race. The Monist League spread throughout Germany and Austria, establishing forty-two branches. It published a periodical for its members and another periodical for its youth movement. The word *monist* came from the philosophy of monism, which holds that nature is all one, meaning that existence has no separate spiritual component.

The Monist League was extremely nationalistic, and many of its adherents fell at the front in World War I. Die-hard monists formed

farms and prevent Ukrainian independence. It also made space for Russian immigation. Most of the confiscated grain was exported.

A "hidden holocaust," the Ukrainian famine was covered up with great success by the Soviet government. To rebut rumors of famine, some influential Europeans, including the socialist George Bernard Shaw, were given a carefully prepared tour. From its early days the "world's first scientific state" imitated Prince Grigori Potemkin, who reportedly built the facades of fake villages in order to impress Catherine the Great with the wealth of his Crimean conquest.

The foreign press in Moscow wasn't duped, but it wasn't reporting much either. The reporters either ignored the horrific events in the Ukraine or fudged their dispatches in order to please the Soviet censors. Journalist Malcolm Muggeridge told the truth, but he did not sign his articles, and they were put on back pages of the *Manchester Guardian*. The best reporting was done by a twenty-eight-year-old Welshman, Gareth Jones. Jones was a Cambridge University graduate fluent in the Russian language, and he on foot made a forty-mile trek through starving villages, later submitting reports from outside the country.

Unfortunately, Jones's first-hand accounts were contradicted by articles in the prestigious *New York Times*. Authoritative-looking rebuttals were contributed by the Pulitzer Prize winning Walter Duranty. He was a reporter with eleven years experience in the Soviet Union, and he was highly respected for having obtained an exclusive interview with Stalin. (Let us not be so eager to praise exclusive interviews with famous men.) Presumably wishing to keep on good terms with Stalin, Duranty asserted that any claim of famine was either an "exaggeration or malignant propaganda." There was hunger, admitted the *Times's* holocaust denier, but no famine, and he excused Stalin's harsh policies by saying, "You can't make an omelet without breaking eggs."

Duranty, who was born British, gave very different information to the British embassy in Moscow. He confidentially advised a diplomat that the famine's death toll possibly amounted to 10 million.

Gareth Jones would not sacrifice his integrity to keep on good terms with the authorities. Accused of espionage, Jones was banned from the Soviet Union and went to China. There he was killed, apparently by bandits, while reporting on the Sino-Japanese War. In the now independent nation of Ukraine, the intrepid, slightly built Gareth Jones is remembered as an "unsung hero." In 2005 the Ukrainian parliament declared the famine a "genocide," and in the following year the Ukrainian president, Viktor Yushchenko, over objections from the Russian government, began a campaign to have the United Nations make a similar declaration.

William Jennings Bryan lost the public relations battle at the Scopes Monkey Trial. But in 1915 he had been correct in perceiving the United States as heading into World War I, even though President Wilson appeared to favor peace, and in 1925 Bryan was correct to predict dire consequences from the acceptance of Darwinism with its "black and brutal" form of evolution.

Chapter Six

A Good Story Is Hard to Kill

Textbooks hoodwink. Cambridge University botanist E. J. H. Corner, *Contemporary Botanical Thought* (edited by Anna M. MacLeod and L.S. Cobley), 1961

The extreme rarity of transitional forms in the fossil record persists as the trade secret of paleontology. Stephen J. Gould, *Natural History,* May, 1977

Why, since 1859, have most paleontologists looked the other way whenever the subject of evolution has been raised? American Museum of Natural History paleontologist Niles Eldredge, *Time Frames,* 1985

In 1872, the giraffe example of evolution entered the sixth and final edition of *The Origin of Species,* and it became the schoolbooks' favorite. In 1987, Stephen Gould examined the major American biology textbooks and discovered that every one of them led off its section on Darwinism by explaining how the giraffe became the world's tallest animal. That explanation had been published for 115 years even though, according to Gould, there was absolutely no evidence for it. "Giraffes," said the professor, "provide no established evidence whatsoever for the mode of evolution of their undeniably useful necks."

According to Darwin (and Wallace ahead of him), Africa's blond giant evolved from some kind of hoofed animal that depended on tree leaves for its sustenance. During times of drought the shorter of the species perished for lack of food while the taller ones survived because they could reach higher on the tree for leaves. Even an "inch or two" of height, said Darwin, could have made the difference between life and death, and so over a long period of time a succession of droughts gradually brought about the giraffe.

Sounds good, doesn't it? In George Bernard Shaw's time, the giraffe was the "central fact in the evolution rumpus," according to his friend and biographer Blanche Patch. It vexed the playwright that, in Shaw's words, there was "no getting away from him."

Gould criticized the giraffe story in a 1988 issue of *Natural History* magazine but to no effect. The Harvard zoologist brought up the subject again in 1996 (in an article titled "The Tallest Tale"), and he wrote about the giraffe a third time in 2000. For long it mattered not to the writers of textbooks that the nation's best known evolutionary biologist condemned their leading example of natural selection as a fantasy for which there was no evidence. Lack of evidence was not the only problem. Here I shall present evidence *against* Darwin's giraffe story.

While traveling aboard the HMS *Beagle*, Darwin visited South Africa, but he did not spend much time there, and he certainly did not learn much about giraffes. In times of drought the principal food of the animal is not tree leaves, as Darwin thought, but the leaves of bushes. Giraffes also can eat grass. It is during the *rainy season*, not during droughts, that the giraffe likes to munch on tree leaves. The leaves of the acacia tree then become succulent and nutritious, and the foliage of an acacia grove is likely to disappear up to a certain height consistent from tree to tree. Probably this odd-looking "browse line" is what gave rise to the belief that giraffes engage in a life and death competition for tree leaves.

Among various publications concerning giraffe feeding habits, I recommend "Winning by a Neck" in *The American Naturalist* of November, 1996. The authors performed field studies in Africa. Having debunked the tree leaf story, they maintained their Darwinian credentials by suggesting that the neck lengthened as a result of "sexual selection": taller males seemed to win the jousts for females by whacking their opponents with their longer necks. How that sort of contest happened to get started among some hooved animals and not others, the authors did not explain.

The fanciful tree leaf story that Gould complained about I have found bolstered at the Smithsonian Institution of Natural History. In 2003 I visited the museum's new Hall of Mammals and encountered there the life-sized model of a giraffe reaching for tree leaves. A sign informed me that the animal's height "was quite an advantage."

True but misleading.

The Smithsonian Institution operates the National Zoo. My curiosity having been aroused, I telephoned the zoo to inquire what they were feeding the giraffes. A helpful information person was about to read me something from a book, but fearing I would be told about acacia leaves, I asked to be put in touch with somebody who was involved with the actual feeding. This gentleman informed me that the National Zoo giraffes are fed pretty much the same food as horses. They eat oats, alfalfa hay, and fruit. Occasionally, for variety they are given some leafy branches such as mulberry or bamboo.

In sum, giraffes do not need acacia leaves in order to survive.

Here are a few more weaknesses in the giraffe story:

There is no fossil record of hoofed mammals gradually evolving into giraffes. Darwinists assume that the giraffe evolved from a somewhat long-necked, horse-like species such as the okapi, but the height difference between the okapi and the giraffe is huge.

There is a survival disadvantage to the giraffe's great height. This species has the highest blood pressure of any mammal. It is so high that when chasing a giraffe in a vehicle for the purpose of capture, it is important to get the animal under control quickly before he succumbs to a heart attack.

The following objection ought to have occurred to Darwin (and even to Shaw): *If an inch or two of height were a matter of life and death, how could there be any female giraffes?* They stand two to three feet shorter than the males.

While preparing this chapter, I visited some schools, where I found four different biology textbooks currently in use, and the giraffe appeared in two of them. The campaign that Gould began twenty years ago is a partial success.

Since we have no other good explanation for the giraffe, what about Lamarck's theory? As evidence for the effect of use and disuse, we could note that the giraffe's tongue is armored with tough little protrusions that enable it to ignore the acacia thorns while grabbing the branches, just as the ostrich is born with callosities where he sits. We have no evidence that the armored tongue is a matter of life and death as Darwinism requires. As for whether *striving* can affect an animal's height, perhaps experimentation could determine that.

Now let's talk about horses. In the nineteenth century London was a city full of horses, and the English love of them was such that tickets sold out when Thomas Huxley lectured on how the horse evolved in Europe. But he was selling false information. In 1876 Huxley came to the United States to lecture and to visit his emigrant sister Eliza, and Othniel Marsh at Yale University showed the English scientist a large array of fossils demonstrating that the horse had not evolved in Europe but in North America. Thoroughly convinced, Huxley reversed himself and began lecturing on the horse's evolution in North America.

It came to be assumed that the modern horse crossed the Bering Land Bridge connecting North America to Siberia, then ambled all the way to Europe, where, oddly, it encountered the pre-horses that Huxley originally lectured about.

For reasons unknown, in both North and South America, the horse became extinct approximately eleven thousand years ago. Possibly the species was killed off and eaten by the people who crossed eastward on the Bering Land Bridge to become the "Native Americans." In colonial times the horse was brought back to the Americas by the Spanish.

As we now know, Marsh's arrangement of fossils erred in showing that over time there was a regular, one by one reduction of toes into a hoof. But Darwin said this was the best example of evolution that he had ever seen, and Marsh's fossil chart became ubiquitous in textbooks, reference books, and museums.

Actually, in the evolution of the horse, approximately two hundred species were involved; the physical changes went in all directions; and the straight-line progression that Marsh published was purely imaginary (like the often pictured straight-line evolution of *Homo sapiens*). It made no difference that Marsh's chart began to be discredited in 1913. Authors loved his example of step by step evolution.

Marsh's fantasy was still going strong in 1951 when the most prominent American paleontologist of the time, George Gaylord Simpson, debunked it in his book *Horses*. This failed to penetrate the consciousness of many writers on that subject, and so Simpson debunked Marsh's fantasy more emphatically in his 1953 book *Major Features of Evolution*. In that book he said, "The most famous of all equid trends, 'gradual reduction of the side toes,' is flatly fictitious."

In a lighter vein Simpson added that there was a widespread tendency to "put the chart before the horse." In true Darwinian tradition, the fossils were arranged to fit the theory, not the theory to fit the fossils.

A fossil display of Marsh's type at the American Museum of Natural History was criticized in a 1988 book, *Darwin's Enigma*. The author, Luther Sunderland, quoted the museum's own Niles Eldredge: "I admit an awful lot of that [imaginary evolution stories] has gotten into textbooks as if it were true. For instance, the most famous example *still on exhibit downstairs* [emphasis added] is the exhibit on horse evolution prepared perhaps fifty years ago. That has been presented as literal truth in textbook after textbook." According to Eldredge, the people who sponsor these erroneous, straight-line displays say they are just letting the fossils speak for themselves. "But the message comes through loud and clear," said Eldredge to Sunderland, "anyone looking at that exhibit is bound to come away with the notion that evolution is a matter of gradual, progressive change through time."

As a typical textbook favorite, Marsh's horse chart continued to be published everywhere, and it still is seen in some recent books almost hundred years after the initial objections.

We have skipped over a few interesting questions. What about all those pre-horse fossils that Huxley lectured about before visiting the United States? How did they happen to be on the eastern side of the Atlantic? The earliest pre-horse fossil, *hyracotherium*, was found in England in 1838, and the same species turned up later in the United States. (The latter was named *eohippus*, but when it became known that the two animals were the same, the term *eohippus* was abandoned in favor of the earlier discovery, *hyracotherium*.) How did *hyracotherium* happen to be in England, which 50 million years ago was an island off the coast of what is now Europe? It has been suggested that pre-horse species spread out from Central Asia, but Simpson found no evidence of that. Neither has anybody identified a fossil ancestor for *hyracotherium*. Yet the *Encyclopaedia Britannica* tells us, "The evolutionary lineage of the horse is among the best documented in all paleontology."

What has been called the "Piltdown moth" is an even stranger saga. Until recently, this fantasy appeared in virtually all the biology textbooks and it is not gone yet. Supposedly the smoky air of England's nineteenth century industrialization caused the evolution of a new, dark colored moth, which was named *Biston carbonaria*. One explanation held that caterpillars were ingesting soot and so producing dark colored moths. We also heard that *carbonaria* was a genetic mutation that happened to proliferate because predatory birds could not see dark moths when they rested during the daytime on soot darkened tree trunks. Old records indicated, however, that dark moths were nothing new. Then we heard about the dark moths as a "recurring mutation."

The Great Moth Fantasy became prominent in the 1950s when the British medical doctor and amateur entomologist Bernard Kettlewell claimed to have proved with experiments that, in polluted areas, newly evolved dark colored moths survived better than light colored moths. For the *Scientific American* issue of March 1959, Kettlewell wrote an article titled, "Darwin's Missing Evidence." The text informed us that if Darwin had noticed what industrialization was doing to moths during his lifetime, "he could have observed evolution in action."

In a scientific journal Kettlewell said the dark moth represented the "evolution of melanism."

As in the case of Piltdown man there soon came objections, but the textbook writers fell in love with the moth story and ran with it. The new fantasy could be presented clearly and simply like the giraffe story and the horse story. Photographs showed how advantageous it was for dark moths to rest on dark trunks and light moths on light trunks; never mind that some researchers found in the wild no correlation with respect to color and survival. Next, anti-pollution legislation in England was claimed to have brought about a declining percentage of dark colored moths. But a similar decline occurred also in parts of Michigan where there never had been any pollution.

Furthermore, there were reports of dark colored moths thriving in the snowy Arctic.

Leaving aside all the back and forth argumentation, the operative fact is that moths *do not voluntarily rest upon tree trunks*. During the daylight hours the little fellows quite sensibly prefer to go and hide somewhere. This fact has been well established since at least 1980. One moth expert reported that in twenty-five years of field work, he saw only one of the tiny creatures land on a tree trunk.

So how did Kettlewell do his research? He said that he had collected dark and light moths and "released" them onto tree trunks. That sounded as if the "released" moths headed like homing pigeons to the nearest tree trunk. But Kettlewell admitted in a scientific journal that his moths were "not free to take up their own choice of resting sites." We do know that later researchers used dead moths and fastened them, wings outspread, to tree trunks with glue or pins. We also know that moths photographed for textbooks have been fastened. *The Scientist* of May 24, 1999, carried a detailed debunking of the moth fantasy. The author, Jonathan Wells, complained that Kettlewell's moth nonsense still was a feature of biology textbooks and "taught to every student of biological evolution."

Scientifically, the whole moth controversy has been a tempest in a teapot when one considers the fact that the light colored moths and the dark colored moths actually are members of the same species, *Biston betularia*. They are just different varieties and can breed with each other. Therefore, *carbonaria* involves no evolution at all, Kettlewell to the contrary. The amount of difference between the light moths and dark moths is no more than the differences involved with dog breeding. Perhaps even less. A Darwinian biologist once told me that the Great Dane and the Chihuahua are different species because the difference in size is so great that they cannot breed. I thought they were both *Canis familiaris*.

Bernard Kettlewell was praised for the elegance and simplicity of his experiment, which was called a "great classic of evolutionary biology." Like Charles Dawson, however, he never was elected to the Royal Society. For whatever they might signify, I shall mention a few other things that Kettlewell and Dawson had in common. Both were notable for their enthusiasm and infectious good spirits. Both worked outside his regular field, and both met a premature end. In 1979 Kettlewell, although a medical doctor, gave himself "an apparently accidental overdose" of pain medication, according to the *Dictionary of Scientific Biography.*

Gould must not have followed the moth controversy closely. When battling the creationists in his 1983 book *Hen's Teeth and Horse's Toes*, he brought up the melanic moth claim as "evidence of evolution in action" saying, "they became black when industrial soot darkened the trees upon which the moths rest. (Moths gain protection from sharp-sighted bird predators by blending into the background.)" With the richest of irony the professor added, "Creationists do not deny these observations; how could they?"

A good Darwinian fable really is hard to kill. In 2008, I visited a large butterfly conservatory, and I learned that the Kettlewell nonsense still was being taught there. It also popped up in the *New Scientist* magazine of April 19, 2008, as a key element in a long article explaining evolution.

Now we come to butterfly mimicry, still a popular topic in the textbooks. Supposedly the non-poisonous viceroy butterfly evolved to look exactly like the poisonous monarch butterfly so that birds would not want to eat it. This is plausible until one thinks about it, as did the geneticist Reginald Punnett back in 1915. His objection: If the predators avoided the viceroy just as it began to change its appearance, their eyesight was not very good. That being the case, how did the viceroy become a perfect mimic? Why not just a 50 percent mimic? Or a 99 percent mimic? How did those weak-sighted birds manage to evolve a 100 percent mimic? (My question: how do we know which butterfly came first?)

Trying to keep up to date, I checked a recent college level textbook to make sure the mimicry story still was being told. It was. But in continuing research on the topic I happened to find out that this classic case of natural selection had been conclusively debunked eighteen years ago. In the journal *Nature* of April 11, 1991, two researchers reported that the viceroy butterflies are as unpalatable as the monarchs and declared the mimicry story false.

Gould's books on natural history had quite a following among the educated public, and he got into trouble with his colleagues by calling attention to evolutionary problems such as the lack of gradualism. Paleontologist Steven M. Stanley in his technical book *The New Evolutionary Timetable* also commented on the problem with gradualism:

> Since the time of Darwin paleontologists have found
> themselves with evidence that conflicts with gradualism,
> yet the message of the fossil record has been ignored . . .
> It was soon forgotten that Darwin's judgment of the fossil
> record was based on deduction rather than fact.

As noted earlier, Darwin insisted that to deny gradualism was "to enter into the realms of miracle." Apparently the kangaroo rat is a miracle, and so is the pocket gopher, and so is the pocket mouse. All three carry food in external cheek pouches. How could such a thing evolve? Gould brought up this question in his book *The Panda's Thumb,* saying he could imagine *internal* cheek pouches evolving little by little as the animals tried to carry more and more food in their mouths. He could not imagine *external* pouches evolving gradually. "The weight of these [the rodents with external pouches], and many similar cases," said Gould, "wore down my faith in gradualism long ago." (Notice how that word *faith* keeps coming up.)

The lack of gradual transition is common in evolutionary history, but some instances are more striking than others. Ernst Mayr listed as problem novelties the "bird feather, the mammalian middle ear, the swimbladder of fishes, the wings of insects, and the sting of aculeate hymenopterans" (wasps and bees).

The feather problem came to mind as I was reading the anti-creationist book *Scientists Confront Creationism*, a collection of writings edited by anthropologist Laurie Godfrey. Godfrey herself wrote a chapter titled, "Creationism and Gaps in the Fossil Record," and so I looked in there to see what she had to say about the bird. It turned out that Godfrey had nothing to say about the bird. She advised me to read a different book, *The Age of Birds*, by the Harvard University ornithologist Alan Feduccia. So I got hold of Feduccia's book, and it said that feathers evolved from the scales of a reptile. But Feduccia stated frankly that there was "no known structure intermediate between scales and feathers." There was a gap that Godfrey had glossed over.

Concerning the feather problem Feduccia referred me to five other authorities, and so I looked them up too. Among those publications, all scientific journals, I found what turned out to be the standard explanation for the feather, which was published originally some eighty years ago by the Danish naturalist Gerhard Heilman. Feduccia himself adopted Heilman's idea in his later book *Origin and Evolution of Birds*. You might find this preposterous, but according to Heilman and now Feduccia, some scaly creatures took to jumping from trees and *wind friction* frayed their scales until they became feathers. (Yes, you read that right. Wind friction.) Another of the authorities cited by Feduccia said that the leaping lizards' bumping into tree branches also might have caused feathers to form, complete with rachis, barbs, barbules, and hooklets. A wonder to behold.

The feather is a complex object of various parts and a marvel of aerodynamic efficiency. The feathered wing is phenomenally strong for its weight; it can open and close slots like an advanced airplane wing; it can change shape as needed. Amazingly, the feathered wing appears suddenly and fully formed in the fossil record. So does the insect wing, another great mystery.

Assuming it possible that wind friction or branch bumping could bring about the feathered wing, it surely would require many generations of leaping lizards to accomplish that. Otherwise one could train an electric fan on a lizard, make feathers, and win a Nobel Prize in physiology. Either way we are talking not about Darwinism but *Lamarckism*. The frayed scales would have to be *inherited* and the inheritance of acquired characteristics is a tenet of Lamarckism, which is ridiculed and despised by Darwinists.

Feduccia's excursion into Lamarckism ought to have caused an uproar among evolutionary biologists. Mysteriously, he seems to have gotten a pass. Perhaps this is because nobody wants to call attention to the feather problem.

Not so lucky was Otto Schindewolf. In the 1930s, this exceptionally erudite German paleontologist got himself roundly castigated by declaring that a reptile laid an egg and a bird came out of it. (A remarkable saltation—or miracle.) That was what the fossil record showed. Nowadays Thomas Huxley's notion that dinosaurs evolved into birds has come back into fashion, and so we could update Schindewolf by saying that a dinosaur laid an egg and a bird came out of it.

A German geneticist, Richard Goldschmidt, supported Schindewolf by theorizing that small genetic changes occurring in the womb could produce large evolutionary jumps. Goldschmidt came to the United States as a refugee from Nazi Germany, which was persecuting Jews. In this country he was persecuted for his saltationism. Published in 1940, his book *The Material Basis of Evolution* said that new species did not come about by the accumulation of tiny changes known as microevolution. Between species, said Goldschmidt, there were "bridgeless gaps."

So much calumny was heaped upon this maverick continental that Gould compared it to how the fictional Emmanuel Goldstein, an "enemy of the people," was abused in George Orwell's novel *1984.* Among biologists the mere mention of Goldschmidt's name would trigger an Orwellian Two Minute Hate, and this was still going on in the 1960s when Gould was a graduate student. Another biologist, Thomas Frazzetta, has been quoted as saying that when he was a student, Goldschmidt was "always introduced as a kind of biological 'in' joke, and all we students laughed and sniggered dutifully to prove that we were not guilty of either ignorance or heresy." Goldschmidt himself marveled at how he "had struck a hornet's nest" and was viewed as "not only crazy but criminal."

Whatever happened to scientific objectivity? Did Thomas Huxley demolish it with his beak, claws, and X Club?

Goldschmidt realized that mutations almost always produced some sort of monster that was unlikely to survive. However, the geneticist figured that once a while some "hopeful monster" *would* survive, as in the case of the feathered one. How that feathered Frankenstein could find a mate was another problem. Yet Goldschmidt's idea has been gaining favor, according to Olivia Judson's *New York Times* article of January 22, 2008, titled "The Monster Is Back and It's Hopeful." Judson explained "A quick survey of nature shows a variety of traits likely to have evolved in one jump, rather than gradually." University of Chicago geneticist Jerry Coyne denounced the article as "silly" and "irresponsible."

Apparently not many biologists realize that the origin of the feather is a mystery. Typical is the following case. In e-mail correspondence with a geneticist (he apparently mistook me for a geneticist because of something that I had published), I mentioned that the origin of the feather was unknown. After consulting with another biologist, he responded that the feather had evolved from scales. That is the answer I always get from scientists, and they always think it is sufficient. I asked the geneticist how scales could have turned into feathers and never heard from him again.

A distinguishing characteristic of mammals is hair, and hair also is assumed to have evolved from scales. Looking through books on mammal evolution, I never found a word as to how that could have happened. I talked to a paleontologist specializing in mammals, and, apparently somewhat embarrassed, he said he had no idea how hair had come about. In further conversation he mentioned that all mammals have hair, although some have very little. The juvenile porpoise has a mustache, which fades away.

If we keep after them, perhaps the evolutionary biologists will find out how hair came about, and that would be a discovery very welcome to the balding portion of our population. Meanwhile, the Smithsonian's Hall of Mammals needs revision. Where at the beginning the exhibit tells us that hair distinguishes the mammals from all other animals, it ought to mention that science has yet to discover how hair came into existence. That, too, might give some emphasis to research on hair.

Meanwhile, let us note that problem novelties normally are followed by vast numbers of associated species. Hence we see many kinds of bird and many kinds of mammal.

In his book *The Material Basis of Evolution* Goldschmidt listed these unexplained novelties: hair, segmentation of arthropods and vertebrates, teeth, shells of mollusks, ectoskeletons, compound eyes, blood circulation, alternation of generations, statocysts, ambulacral system of echinoderms, pedicellaria of the echinoderms, enidocysts, poison apparatus of snakes, whalebone, and primary chemical differences like hemoglobin v. hemocyanin. According to Goldschmidt, "Corresponding examples from plants could be added." Elsewhere, I have seen the nourishing amniotic egg, such as the chicken egg, listed as a major evolutionary mystery.

Inexplicably appearing life forms are the rule not the exception. The non-feathered pterodactyl had long been extinct when bats materialized in the fossil record fifty-five million years ago. The bat emerged with no known evolutionary ancestors; yet recent research shows that this little sonar-equipped mammal is a better flyer than the bird in terms of maneuverability and energy efficiency. (Let's hear it for us mammals.)

Encyclopaedia Britannica has reported that early fishlike fossils are "too fragmentary to permit tracing the modern fishes precisely to their origins." From the *Americana* we have received this bombshell:

"Especially puzzling is the fact that all eight classes [of fish] first appeared in the fossil record almost simultaneously and as recognizably distinct groups." In other words, there was a fish explosion and the major types apparently emerged independently of each other. Such an event Darwin thought fatal to his theory. Said he in *The Origin of Species:*

> On the sudden Appearance of whole Groups of allied Species
>
> If numerous species, belonging to the same genera or families, have really started into life at once, the fact would be fatal to the theory of evolution by natural selection. For the development by this means of a group of forms, all of which are descended from some one progenitor, must have been an extremely slow process; and the progenitors must have lived long before their modified descendants.

The intelligent design people might wish to claim that each unexplained development marks an intervention by a higher power or at least evidence of an intelligent first cause.

In 1980, criticism of Darwinism turned up at the British Museum of Natural History, which was founded as a citadel of Darwinism, and the institution was taken to task for that by the *Nature* scientific journal, which also was founded as a citadel of Darwinism and was still operating the way the X Club intended. The uproar began with the celebration of the museum's centennial. Evolution was the subject of a new publication in which a certain controversial sentence started out, "If the theory of evolution is true . . ." As another egregious offense, an exhibit at the museum displayed organisms in a cladistic manner. This means that the exhibit primarily showed how organisms were *related* to each other instead of showing how organisms had *evolved*.

Cladism was a fast-growing discipline in paleontology. Many fossil experts had come to realize that it was virtually impossible in the vastness of geologic time and the wilderness of life forms to establish which species had evolved into which species. These guesses turned out to be wrong. For example, there was worldwide excitement when the supposedly extinct coelacanth was discovered alive in South African waters, and this lobe-finned fish was trumpeted as the evolutionary ancestor of land animals—including us. But now it is not. Gould once gleefully reported that the "smoking gun" had been found concerning evolution of the whale. Not long afterward, that "smoking gun" was discarded and another put in its place.

In response to the centennial publication and exhibit, *Nature* published extremely critical letters as well as blistering editorials. Because of such pressure the exhibit was revised. But twenty-two of the museum's biologists stood their ground even though, as the most alarming charge, in this era of the Cold War, they were accused of promoting a "fundamentally Marxist view of the history of life." Professor L. B. Halstead viewed cladism as the enemy of gradualism, and he pointed out that in Marxist philosophy qualitative changes occur abruptly rather than gradually—as when revolutions take place. Therefore, should the cladists succeed in discrediting gradualism, Halstead warned, Marxism would be able to call upon the scientific laws of history in its support. (Of course, Marxism already was calling upon natural selection theory in its support.) The twenty-two biologists signed a letter to *Nature*, saying that Darwin's theory was supported by "overwhelming circumstantial evidence," but that it ought not to be presented as fact. "The theory would be abandoned tomorrow," said the letter, "if a better theory appeared."

Causing my jaw to drop, *Nature* objected to the biologists' attitude of "agnosticism." Said an editorial:

> The trouble with agnosticism is that, however well justified logically, it can be carried too far. If young scientists are brought up to believe that all theories are falsifiable or metaphysical, is there not a serious risk that regard for the scientific enterprise as a means of understanding how the world is made will be undermined?

A century after Darwin, the shoe was on the other foot. What could be more hypocritical than an evolutionist criticizing agnosticism and accusing a purely scientific exhibit of corrupting the youth? A demand for *faith* was coming from one of the world's most eminent scientific periodicals.

The author of the controversial centennial publication, paleontologist Colin Patterson, contributed a separate letter to *Nature* in which he stated that cladistics is just a method of analysis and not really about evolution. That must have angered Halstead too. The professor wanted the museum staff to apply themselves to proving that Species A gradually evolved into Species B which gradually evolved into Species C and so forth. Halstead was not eager to know how species related to each other. He wanted a list of *begats*—like in the Bible.

Patterson later made a visit to the United States and there he kicked up another storm. Lecturing at the American Museum of Natural History he said

> Question is, can you tell me anything you know about evolution, any one thing that is true? I tried that question on the geology staff at the Field Museum of Science and Natural History, and the only answer I got was silence. I tried it on the members of the Evolutionary Morphology Seminar at the University of Chicago, a very prestigious body of evolutionists, and all I got there was silence for a long time, and eventually one person said, "I do know one thing. It ought not to be taught in high school."

Patterson was confessing the same ignorance about evolution that had been expressed thirty-some years earlier by the paleontologist William Scott, descendant of Benjamin Franklin.

From New York Darwinists circulated an unofficial transcript of what their radical colleague had said, and the English maverick was deluged with irate mail, especially from scientists in academic positions. Many academics have to teach and answer questions, and they do not like to see the Darwinian ground cut out from under them. The basic charge: a disloyal soldier was giving ammunition to the creationist enemy. Patterson, as he put it, "went through merry hell for about a year." He observed ruefully, "One has to live with one's colleagues. They hold the theory very dear."

In 2008, there still was plenty of Darwinian militance in London. The Royal Society's Director of Education, biologist Michael Reiss, had to resign his position because he said in a speech that students with creationist views should be treated respectfully and allowed to ask questions about evolutionary theory. Reiss said nothing *wrong*, the Royal Society admitted, but his views were misinterpreted in the media as favoring the teaching of creationism, and that embarrassed the society. So Reiss had to go. The Royal Society was not living up to its motto: "On the word of no one."

On neither side of the Atlantic do the Darwinists want people, especially students, criticizing their theory. This is quite different from the situation in other classes, such as history, where teachers usually are pleased to see students take an interest in the subject.

Not only scientists hold the theory dear but even many non-scientists. In 2005 authorities at the Smithsonian Institution tried to get a scientist fired from the National Institutes of Health because, as managing editor of a scientific journal, he had published a paper on intelligent design, and in order to show that this sort of Darwinian intolerance was nothing new, I sent a letter to a newspaper telling of the Patterson case and others. It is gratifying when one can get a letter printed, but I wish the editor had not sabotaged it by inserting "Mr." before Patterson's name. Apparently he wished to insinuate that any person who questioned Darwin could not possibly hold a doctorate, even if he were, as I said, a "British Museum paleontologist."

An even more prominent British skeptic was E. J. H. Corner, professor of tropical botany at Cambridge University. Corner was the recipient of the Darwin Medal (1960), the Linnean Gold Medal (1970), and the Japanese International Prize for Biology (1985). The reader might have noticed that most public discussion about evolution refers to animals. Corner was a world class expert on plants, and in the book *Contemporary Botanical Thought*, he told of how Darwinism—a "romantic vision," as he called it—did not explain well the plant kingdom. "I still think that to the unprejudiced," said he, "the fossil record of plants is in favor of special creation."

Interestingly, Corner did not question the existence of gradualism. To him the development of the eight hundred species of *Ficus* was so gradual that it was hard to draw lines between them. But, said Corner, their development occurred in such an even manner, and with so many parallel developments, as to belie the action of randomness combined with natural selection. (Haeckel, though not a creationist, would have understood that.) Indeed, the tropical *Ficus* species, the Cambridge man said, were so prolific and adaptive that they tended to construct their own environments. He also pointed out that the old species of *Ficus* survived along with the new. Corner perceived directiveness in the proliferation of this amazing genus.

One of Corner's contemporaries at Cambridge, E. S. Russell, wrote a book about directiveness, which he viewed as an irreducible characteristic of life. Russell gave many examples of goal-directed activity. One was the healing of wounds; he recounted in awesome detail the complex processes involved. Russell also told of how sometimes a small part of an organism can reconstitute the whole. If one squeezes just part of a living sponge through a piece of bolted silk, he said, the disassociated cells will get together again and create a whole new sponge with flagellated chambers, skeleton, mesenchyme, and other tissues.

Heroic in his own way is the planaria, the tiny flatworm familiar to biology students. If cut in half, the flatworm will reconstitute itself into two complete animals. He will do this, using material and energy from his own body even if he is in a condition of starvation. Often the goal is seen to be more important than survival. The mother bird will risk her life to save her chicks—and sometimes even another bird's chicks.

The concept of directiveness carries us back to Lamarck and his belief that nature has a self-perfecting tendency, which today might be called intelligent design.

Chapter Seven

Experts Do the Math

These objections to current neo-Darwinian theory are very widely held among biologists generally, and we must on no account, I think, make light of them. Nobel Prize winning physiologist Sir Peter Medawar, chairman of the Wistar Conference, 1966

In the first part of the twentieth century evolutionary theory was in chaos, but the Anglo-American biologists went to work on that and came up with a revised theory that they called "neo-Darwinism." (It also was called the" modern synthesis," a name taken from Julian Huxley's 1942 book *Evolution: The Modern Synthesis.*) Previously there had been a different kind of neo-Darwinism, but we can skip that. The new neo-Darwinism became the standard textbook Darwinism at least for the English-speaking world.

The French favored Lamarck, and the Soviet Union adopted a politically correct form of Lamarckism known as Lysenkoism, which, on the basis of false claims, helped to ruin Soviet agriculture. In a *Nature* issue of around 1980 I read that the French still favored Lamarck, and in a 1983 book on the history of Darwinism Peter J. Bowler reported, "Even today it is said that the majority of French biologists refuse to take the modern synthesis of genetics and selection theory seriously." In the early 1990s I wondered whether the French were still Lamarckian. An internationally minded American paleontologist advised me, "Yes, and their theory is no better than ours."

The most distinguished French evolutionary biologist of the twentieth century was Pierre-Paul Grasse, who. as mentioned axearlier, derided Darwinism as a "pseudo-science." Grasse held the Lamarckian belief that evolution basically was a self-perfecting process driven by some internal factor.

The trickiest part of neo-Darwinism is called "population genetics." This purports to show by mathematics that Mendel's laws of inheritance do not conflict with gradual evolution by natural selection. The calculations were done by two British biologists, R. A. Fisher and J. B. S. Haldane, and one American biologist, Sewall Wright. In neo-Darwinian theory evolution proceeds by recombination of genes and the accumulation of tiny multitudinous genetic mutations rather than by single mutations big enough to create new species. Without mutations, it is assumed, the evolutionary process would run out of genetic variation. Darwin would have been pleased to know that the neo-Darwinian process in theory is extremely gradual and in time large changes result. Unlike Darwin, the neo-Darwinists reject the inheritance of acquired characteristics (although as we have discussed, this has been claimed by ornithologists to have occurred when leaping lizards frictionalized into birds).

Neo-Darwinism measures fitness in a very simple way: by the number of offspring. *The more offspring, the fitter.* Shielded by that doctrine, no longer does the biologist have to explain how the giraffe grew so tall or where the feather came from. *A large number of offspring is assumed to represent fitness, and that is all we need to know.* Hearing, that Samuel Butler might have responded, "Humbug!"

After having taken refuge in mathematics, the Darwinian biologists came under attack from mathematicians. Actually, even in the time of Charles Darwin at least one mathematician viewed as impossible evolution by the combination of random variation and natural selection. This was the Englishman Alfred W. Bennett, who in 1870 published a paper titled, "The Theory of Natural Selection from a Mathematical Point of View." Bennett contended that Darwin's evolution by insensibly fine steps demanded so many chance variations in a certain direction as to make an improved species mathematically impossible.

Ahead of Punnett, he took up the question of mimicry in butterflies. Assuming that a thousand variations would be needed to produce a mimic, probably an accumulation of twenty variations, Bennett decided, would be needed to achieve some protective effect and that being the case, the first twenty would have to result entirely from chance (that is, without the guiding effect of natural selection). Then came this calculation:

Suppose there are twenty different ways in which a *Leptalis* [butterfly] may vary, one only of these being in the direction ultimately required, the chance of any individual producing a descendant which will take its place in the succeeding generation is 1/20; the chance of this operation being repeated in the same direction in the second generation is . . . 1/400; the chance of this occurring for *ten* successive generations . . . is about one in ten billions.

The odds were such that random variations would make no significant progress toward mimicry.

Charles Darwin was not much interested in mathematics. While a student at Cambridge, he "attempted mathematics," according to his autobiography, and found the subject "repugnant." In the summer of 1828 Darwin hired a tutor (a "dull man") to help with basic algebra, but the student "got on very slowly," and that summer marked the end of his mathematical endeavors. When writing his autobiography, Darwin "deeply regretted" not having learned more about mathematics because "men thus endowed seem to have an extra sense."

Probably Darwin's lack of facility with numbers was not genetic but psychological. His son George, the sickly one whose health improved after he married an American, was a mathematician who became a professor of astronomy at Cambridge.

As mentioned in the prologue, in the early twentieth century Lecomte du Nuoy launched a mathematical assault on abiogenesis.

In the 1960s, the mathematicians acquired mainframe computers, and some of them, having taken a look at neo-Darwinian theory, reached the conclusion that random genetic events combined with the demands of survival could not possibly account for the enormous complexity of life.

Informal debates began in Switzerland at a picnic attended by two biologists and four mathematicians. The mathematicians were appalled to see what the biologists thought could be achieved by chance. According to biologist Martin Kaplan, a second picnic featured "several hours of heated debate." The biologists then "proposed that a symposium be arranged to consider the points of dispute more systematically, and with a more powerful array of biologists." While it might look as though the biologists were getting the worst of it, it is to their credit that they wished to deploy their biggest guns and really fight it out.

And so a big international conference was held in 1966 at Philadelphia's Wistar Institute of Anatomy and Biology. On the Darwinian, side one of the biggest and most active guns was Ernst Mayr, who was one of the constructors of neo-Darwinism, though not of the mathematical part. Mayr's original specialty was ornithology, and according to him, observing the great variety of tropical birds made him a Darwinist (despite the feather problem). Of the biologists who had worked out the mathematics for the synthesis, in 1966 two were deceased and the third, Sewall Wright, did not attend the Wistar conference. Wright's absence tended to defeat the purpose of this unprecedented and historic event, but he submitted a paper and the conference participants made references to his work.

The moth man, Bernard Kettlewell, was there, but, for whatever reason, he did not submit a paper. Mayr called attention to his work. The Harvard man noted smoothly: "With Kettlewell in the audience, I hardly need remind you of the extraordinary evolution rates that were exhibited in the establishment of industrial melanism." Mayr made his point without going into the messy details.

Chairman of the conference was a Nobel Prize winning physiologist, Sir Peter Medawar. One would expect a man of his specialty to be on the side of the biologists, but his opening remarks were so critical of neo-Darwinism that one must wonder how he got to be chairman. I suppose the Nobel Prize did it. (By the way, there is no Nobel Prize for biology as such. Otherwise, as some of his colleagues have said, Mayr would have received one.) Opening the conference Medawar remarked, "These objections to current neo-Darwinian theory are very widely held among biologists generally, and we must on no account, I think, make light of them."

"Biologists generally" were objecting to textbook Darwinism? What a news bulletin. But as in the case of the proverbial tree falling in the forest, the public would not hear, and so it fell in silence.

As one objection, Medawar mentioned the criticism that neo-Darwinism explains too much. No matter what amazing thing might have happened in evolution, the theory explains it merely by saying either that genes recombined or that mutations took place. In short, whatever happened the genes did it. End of story. Such an explanation might be true, but it sheds little light, and critics were calling neo-Darwinism "vacuous."

As in Switzerland the debate was heated. This delighted the mathematician Stanislaw M. Ulam, who for the first time in his career was experiencing a scientific meeting that was not boring but full of passion, just like the meetings which, as a boy, he had read about in Sir Arthur Conan Doyle's book, *The Lost World*. "It is all very wonderful," he enthused.

The Lost World is the book, it will be recalled, whose sales benefitted from the Piltdown excitement. Doyle's main character, Professor Challenger, was prone to violence and had a terrible temper that enlivened any meeting he attended. Quite possibly Doyle got the idea from the two-fisted fossil hunter Edward Drinker Cope. Cope was a famous American paleontologist who, although reared as a Quaker, was so aggressive that he even got into a fistfight at the American Philosophical Society. Cope's fossil collecting rivalry with Othniel Marsh was so full of spying and dirty tricks it has been called the "bone wars."

Leading off the mathematicians' attack was Murray Eden of the Massachusetts Institute of Technology, who titled his paper, "Inadequacies of neo-Darwinian Evolution as a Scientific Theory." The paper offered several arguments that the formation by chance of a living organism or even a complex protein was mathematically impossible in the time our planet has existed. Eden brought up the very complex protein hemoglobin, which in order to function must be fully assembled. (Mivart would have asked, what good is half a hemoglobin?) Biologist George Wald from Harvard University backed up Eden's contention by commenting that only one mutation in hemoglobin is enough to ruin the blood and kill the organism.

To digress for a moment, Wald in the 1950s used to give a lecture on the origin of life in which he said that life is so complex that its origin from simple chemicals must have consumed an immense amount of time. Wald wrote in the previously mentioned *Scientific American* article: "The time with which we have to deal is the order of two billion years . . . Given so much time, the 'impossible' becomes possible, the possible probable, and the probable virtually certain." This was the sort of plausible guesswork to which the mathematicians were objecting. Furthermore, since Wald published that explanation for the origin of life, the two billion years have virtually disappeared. Evidence for life now has been traced back 3.85 billion years, and as Gould has said, "Life probably arose as soon as the earth became cool enough to support it." There went a once very persuasive argument.

In a paper which was submitted in response to Eden's, the absentee Sewall Wright questioned Eden's understanding of the evolutionary process and said that natural selection did not require so many "operations" as Eden thought. As "not a perfect analogy" to natural selection, he mentioned the game Twenty Questions, which, said

Wright, was "enormously more like natural selection than the typing at random of a library of 1,000 volumes with its infinitesimal chance of arriving at any sensible result" (an Eden analogy).

Of course, in Twenty Questions people try to ask logical questions, not random ones, but I no doubt missed something in Wright's abstruse argument.

Marcel Shutzenberger, a mathematician from the University of Paris, asserted that computers now were capable of calculating the probabilities of evolutionary theory, and as a result, the theory was shown not to work. "I want to know," said the mathematician, "how I can build on computers programs in which—"

"We are not interested in your computers!" barked the British geneticist C. H. Waddington.

"I am!" responded Schutzenberger.

Schutzenberger was accused of arguing for creationism. "No!" came a chorus of response, presumably from the mathematicians.

Wald, who would receive his Nobel Prize in physiology in the following year, reported as an example of natural selection how the retinas of fish eyes are adapted to the different wavelengths of light available at different depths in the sea. Fish near the surface have one type of retina, fish at a 200-meter depth have another, and those swimming completely beyond the reach of sunlight have a retina sensitive to the wavelength given off by luminescent bacteria that cling to certain fish. To Wald those were examples of natural selection. Lamarck and both Darwins might have said they were examples of use and disuse.

Wald called the fish and bacteria relationship symbiotic since the bacteria provide a faint glow of illumination while the fish provide a habitat for the bacteria. However, the advantage to the fish became obscure as the matter was discussed. According to Wald, the bacteria provide so little light that "very, very rarely" would it be useful. And as Mayr pointed out, the luminescence would make the fish more visible to predators. Wald remarked that if the luminescent fish lived in schools, that would make the glow "beautifully useful," but he said, "I suspect that they are more often solitary."

So Wald's argument for natural selection fizzled, and the biologists must have been wishing that he had stayed home.

Schutzenberger asked why a certain European eel has not evolved, like other eels, so that it could breed in its home waters. Instead, for millions of years this eel has made a long and "dangerous" trip to the Sargasso Sea and back.

"Maybe they are having a hell of a good time; how do you know?" was Mayr's retort.

Until his death in 1996 Schutzenberger continued his battle against neo-Darwinism, according to a letter sent to me by his son. Murray Eden I talked to long after the conference. He thought the biologists did not understand the mathematical papers.

Neo-Darwinists doubted that the mathematicians understood evolution. For that matter, how many non-specialists in the field understand modern evolutionary theory? While most scientists (like Carl Sagan) strongly support Darwinism, few, not being evolutionary biologists, know more than what they have learned from their flawed textbooks on general biology.

Waddington declared as "outdated and old fashioned" the commonly held concept of fitness. He said that fitness was measured by the number of the organism's offspring, not by the Darwinian concern for how fast a horse could run or how well a moth could hide. "Nothing else [than the number of offspring] is measured in the mathematical theory of neo-Darwinism," said Waddington.

He went on, "It is smuggled in and everybody has in the back of his mind that the animals that leave the largest number of offspring are going to be those that are best adapted [for survival], *but this is not explicit in the theory"* (emphasis added). Waddington admitted as "vacuous" the neo-Darwinian principle that, as he put it: "Natural selection is that some things leave more offspring than others; and you ask which leave more offspring than others; and it is those than leave more offspring." There is "nothing more to it than that," said he.

As the antiwar poet Sarah N. Cleghorn observed, "The unfit die—the fit both live and thrive. Alas, who say so, they who do survive."

Most evolutionary biologists are geneticists. A relatively small number are paleontologists, and they are the ones who try to figure out how organisms are affected by their environments. Their work is not vacuous, Waddington acknowledged. But outdated as he said earlier. According to the geneticist John Maynard Smith, the geneticists are inclined to tell the paleontologists "to go away and find another fossil, and not to bother the grown-ups."

In other words, the geneticists are not really concerned about how the environment uses natural selection to design new organisms; they merely "smuggle in" the idea in order to account for the antics of the genes. As for the paleontologists, they have come to refocus much of their effort toward cladism, the study of relationships rather than evolution. They also study the behavior and characteristics of prehistoric organisms. Nevertheless, some representatives of both groups defend textbook Darwinism and ridicule those who are critical of it.

Being numerous, the geneticists tend to dominate the Society for the Study of Evolution, which with its scientific prestige can influence how evolutionary theory is taught. The society battles against consideration of intelligent design in the classroom, even though, ironically, in their laboratories the geneticists themselves practice intelligent design all the time. They call it "bioengineering."

Always aggressive, the German-born Mayr turned the tables on the mathematicians by submitting a paper titled, "Evolutionary Challenges to the Mathematical Interpretation of Evolution." In various ways he showed that the speed of evolutionary change is extremely variable and depends on a variety of circumstances. One species of blue-green alga, for example, has not changed in 900 million years while house sparrows brought to North America in the middle of the nineteenth century already have diversified into various subspecies.

As mentioned earlier, Mayr also brought up industrial melanization as an example of rapid evolution.

The purpose of Mayr's paper was to show that evolution "again and again has resulted in unique phenomena and in startlingly unpredictable phenomena." For that reason, he said, mathematical programs cannot arrive at a realistic interpretation of evolution if they are set up in "too deterministic a manner."

But if evolutionary change is so unpredictable, how do we know that we have the right theory? Darwin admitted to Asa Gray that his theory lacked sufficient fossil evidence but argued that it offered much explanatory power. Mayr weakened that argument.

Eden welcomed Mayr's challenge to mathematics and said that, in his view, some of the problems Mayr mentioned "are, indeed, amenable of mathematical treatment."

Not being a mathematician, I have given only a minimal and perhaps unfair idea of what took place. Anybody who wishes to study for himself the Wistar proceedings can find the papers and discussions in a book, *Mathematical Challenges to the neo-Darwinian Interpretation of Evolution*, edited by Paul S. Moorehead and Martin M. Kaplan.

One of the Wistar participants, the geneticist Richard Lewontin, elsewhere gave a lecture on neo-Darwinism that was attended by the skeptical cell biologist Lynn Margulis. Margulis is noted for her work on the theory of symbiogenesis, which tells us that organisms evolve by combining with each other (the principle expounded, roughly, by Empedocles), especially by the transfer of genetic information from one microorganism to another. The cell with a nucleus now is believed to have originated when one bacterium entered another bacterium and managed to take up residence there. As another example of colonization, the human gut is said to contain 2.2 pounds of bacteria many of which we humans need for the digestion of food.

Lewontin's lecture, according to Margulis (in *The Third Culture*), presented an "elaborate cost-benefit mathematical treatment . . . devoid of chemistry and biology," and at the end of the lecture Lewontin "said that none of the consequences of the details of his analysis had been shown empirically." In other words, there was no proof of the theory's validity.

Why then did he want to "teach this nonsense?" asked the cell biologist.

Lewontin gave two reasons. The first was "physics envy." Biologists yearn for the mathematical certainty available to physicists. The second reason: Neo-Darwinian proposals get the grants.

Medawar brought up the grant problem in his book *Aristotle to Zoos*, a philosophical dictionary of biology. In it he remarked that "neo-Darwinism would be completely undermined by improving learning ability by selection of inbred mice throughout an experiment lasting many generations on the same diet." Medawar concluded pessimistically, "Even if any biologist were willing to undertake such an experiment, no grant giving body known to science would be willing to fund it."

Subscribing to a fashionable theory not only can be profitable but helps in getting published. When Margulis wrote her first paper on nucleated cells it was rejected by more than a dozen scientific journals. But after it finally appeared in *The Journal of Theoretical Biology*, the paper elicited a stunning eight hundred requests for reprints. Margulis's book *The Origin of Eukaryotic* [nucleated] *Cells,* was rejected by the intended publisher even though she had a contract. The book was brought out in 1993 by W. H. Freeman.

Six years after the Wistar conference, the old gradualism problem reared its ugly head with a vengeance. Stephen Gould and Niles Eldredge published a paper on "punctuated equilibrium," a view of the fossil record that conflicted with Darwin's insensibly fine steps.

Whatever one says about punctuated equilibrium, it is bound to be criticized as inaccurate, as if one were trying to count angels on the head of a pin. But here goes. Looking at the fossil record Gould and Eldredge perceived evolution as having occurred in periods of rapid change that featured abrupt emergences of new species with gaps between the new and the old. In addition to those non-Darwinian phenomena, between the periods of rapid change came very long periods in which, mysteriously, no significant change took place. Species, instead of evolving, just *varied around the average type.* (On the Galapagos droughts increase the proportion of finches with big beaks. Wet years favor the small beaks.)

As one cause of no significant change, when the environment turns unfavorable, the most common reaction by a species is not to evolve but to go somewhere else. When the going gets tough, the tough get going. Scientifically that is called "habitat tracking." Henry Fairfield Osborn's wild pig, it will be recalled, roamed between areas that now are Paraguay and Nebraska. But habitat tracking cannot not explain most stasis, which is typical of the fossil record. Consider Mayr's 900-million-year-old alga and the 500-million-year-old invertebrate *Lingula*, another of many "living fossils."

Gould and Eldredge's stop-and-go evolution was not welcomed by most evolutionary biologists. "Evolution by jerks" punned a British geneticist, J. R. G. Turner. There were accusations of saltationism. Some critics viewed punctuated equilibrium with its relatively sudden changes as a Marxist doctrine—like cladism. Suspiciously, Gould's father was known to have been a Marxist, and the Soviet Union's scientists were known to prefer abrupt transitions analogous to the violent social changes that the Marxists so dearly love.

Of course, the notion of sudden change in life forms was nothing new or necessarily political. Thomas Huxley had believed in it. Before him, in the early nineteenth century, the French Baron Georges Cuvier decided that biological changes were sudden and that they resulted from catastrophes. Agreeing with him was Darwin's geology professor, Adam Sedgwick, and, we might as well add, the science-minded George Campbell, eighth duke of Argyll (no Marxist, he). Cuvier believed in "calamities which, at their commencement, have perhaps moved and overturned to a great depth the entire outer crust of the globe." According to the baron, "numberless living things have been the victims of these catastrophes . . . Their races have even become extinct."

Before *The Origin of Species,* catastrophism was a commonly accepted explanation for extinctions, and this fit the Biblical report of a Great Flood. As we shall discuss later, catastrophism recently has returned in the theory of mass extinctions caused by the impact of extraterrestrial objects.

Cuvier's theory was opposed by Sir Charles Lyell, who believed that the geological processes that we see working today are what very gradually produced the world in its present form. Both Lyell and his friend Darwin were dyed-in-the-wool gradualists.

Gradualism won out among geologists to such an extent that an academic career could be jeopardized by finding evidence of a sudden catastrophe. According to David Raup, when the geologist J. Harlen Bretz discovered in the state of Washington the effects of a catastrophic flood caused by glacial melting, the hostile reaction was such that Bretz might have been "drummed out of science" if he had not held a tenured academic position. But in 1979, Bretz, having been vindicated by further research, received a medal from the Geological Society of America. Then ninety-six years of age, he mourned, "All my enemies are dead, so I have no one to gloat over."

As one might expect, Gould and Eldredge's observation of a stop and go pattern in the fossil record had been noticed before—indeed, all the way back to when the word *scientist* was invented. When in 1833 the polymath William Whewell coined that word, a Scottish paleontologist was digging up mammal fossils in the foothills of the Himalayas, and he, although not an evolutionist, noticed that long periods of no change were followed by the sudden appearances of new species.

The Scot, Hugh Falconer, was no obscure scientist. He won the Wollaston Medal, the British Geological Society's highest prize. Huxley needed him to second Darwin's nomination for the Royal Society's Copley Medal, and when Falconer died, he was vice president of the Royal Society. Among other achievements, the Scottish paleontologist, also a medical doctor, found Stone Age tools that proved humankind's existence during prehistoric times.

Despite Falconer's observation and the physical appearance of the fossil record, Darwin's doctrine of gradual, continual change became an unchallengeable doctrine, a great triumph of theory over fact. How did this obvious truth get lost and not reappear until 1972?

Gould's partner, Niles Eldredge, found out why. When a graduate student Eldredge noticed a lack of published evidence for gradualism, and so for his research project he decided to provide some. Basically, however, what he found, as the result of a lot of hard work, was a *lack of change*. In time he came to realize that students tended to avoid this sort of research because if they reported no change, their projects were considered to have *failed*. One does not get a Ph.D. for accomplishing nothing.

In sum, the strong bias toward uniformitarianism in geology was mirrored in evolutionary biology.

Undeterred by disapproval of punctuated equilibrium, Gould wrote another inflammatory paper. In the journal *Paleobiology* he pronounced Mayr's cherished neo-Darwinism "effectively dead, despite its persistence as textbook orthodoxy."

In 1993, the Gould-Eldredge paper's twenty-first anniversary, the uproar had died down substantially, and Gould was allowed to assert in *Nature* the triumph of punctuated equilibrium. But there remained many a dissenter. In 1993 I attended a lecture in which a paleontologist called Gould an "idiot." Just a few years ago a paleontologist told me that punctuated equilibrium was good only to joke about while drinking beer with one's colleagues. He was surprised to hear about the *Nature* article.

By 1999, Gould had become president of the American Association for the Advancement of Science, but Robin Wright reported in the *New Yorker*: "The evolutionary biologists with whom I have discussed his [Gould's] work tend to see him as a man whose ideas are so confused as to be hardly worth bothering with, but as one who should not be publicly criticized because he is at least on our side against the creationists."

The Darwinists of whatever stripe are very concerned about the public's perception of textbook Darwinism, which constantly is under attack from the creationists and intelligent design people. When Gould campaigned for removing the giraffe from textbooks, his stated motive was to keep the public from finding out what a poor example it was. Said the professor:

> If we choose a weak and foolish speculation as a primary
> textbook illustration . . . then we are in for trouble—as
> critics properly nail the particular weakness and then
> assume that the whole theory must be in danger if
> supporters choose such a fatuous case as a primary
> illustration.

Well said.

Despite Gould's efforts, the "fatuous case" continued to dominate textbooks for many more years. As for why, Gould perceived two reasons: the love of a good story and respect for authority (such as *The Origin of Species* and previous textbooks). I would suggest a third reason: Neither Gould nor anybody else provided a good substitute for the giraffe. It appears that except for minor physical changes—as we see among breeds of dog—cogent examples of gradual change are hard to come by. What was the textbook writer to do?

Small changes are called "microevolution," and they can be important to humanity in the realm of bacteria. With a slight change a harmful bacterium reportedly can acquire immunity to an antibiotic. I have to wonder, though, if it is so hard to tell what is a new moth, how can we be so sure of what is a new bacterium? And is microevolution really evolution? Or is it just varying around the average type?

The most striking lack of gradualism occurred approximately 550 million years ago in the Cambrian period. Within a relatively short time emerged all but possibly one of the major animal groupings (phyla) that we have today, and as yet we have no fossil trails leading to them. (The possibly non-Cambrian group comprises the tiny bryozoans; fossils for them have yet to be found as early as the Cambrian.) Arthropods, echinoderms, and mollusks originated in the Cambrian. Of special interest to humankind are the chordates. Theoretically, a possible ancestor of humankind's was the Cambrian period's worm-like pikaia, a 1.5 inch long swimmer that sported the precursor of our spinal cord. Its origin is unknown.

As a "well documented" fossil trail that leads from one group of animals to another, Gould cited the transition from reptile to mammal. There indeed have been discovered steps in the transition. Again we encounter a Scottish paleontologist who also was a medical doctor. This was Robert Broom, and he did extensive work on the reptile to mammal transition. According to the *Dictionary of Scientific Biography*, Broom's "talent for combining paleontological and embryological research enabled him to contribute more to the story of mammalian origins than all his contemporaries together."

Broom, however, was puzzled as to *why* the transition took place. He could find no reason for it. Of course, we humans think mammals are an improvement over reptiles, but did each transitional stage actually offer an advantage for survival? Broom could not see that. The Scot thought Lamarckism was a better theory than Darwinism, but he felt that neither theory was satisfactory. He suggested that "spiritual agencies" were behind evolution. Broom also decided that evolution had come to an end with the creation of the human being. Animals now were too specialized for further evolution, said Broom.

Grasse, too, had thoughts on the transition from reptile to mammal. He asserted that the transition was too direct and involved too many lines of animal, with no apparent pathology along the way, for it to have resulted from the rough-and-tumble of mutation and natural selection. E. J. H. Corner's *Ficus* species come to mind.

Broom claimed that there were hundreds of examples of evolution that could not be explained by Darwinian theory. Following are some that he listed.

The sting of the bee. Darwin said that the sting of the bee helps the species to survive though not the individual. "But the sting is a modified ovipositor [tool for placing eggs]," said Broom, "and we have only to think of the perhaps hundreds of thousands of years during which the organ was neither a useful ovipositor nor a useful sting and was gradually developing into a sting, to see the difficulties." As in the alleged evolution of mimicry, Darwinism always has trouble explaining how new organs evolve by "slight, successive modifications" during the lengthy time in which the modifications are not useful. As an example of that Murray Eden cited complex proteins such as hemoglobin.

Electric fish. A similar problem comes up with the electric fish. Said Broom: "In most fishes the electric organ is a transformed muscle. But we have only to think of the countless generations in which it was neither a useful muscle nor a useful electric organ to see the difficulties of the Darwinian explanation."

Poisonous snakes. "The harmless snakes are far more numerous in species," said Broom, "and in many regions more numerous in individuals than the venomous; so that the development of poison glands and grooved fangs seems unnecessary to preserve their lives." The Scot also questioned the vipers' need for tubed fangs "as beautifully developed as a hypodermic needle." Broom asked, "Must we believe that the ancestral vipers, which only had grooved fangs like the cobras, had to die out in competition with those snakes that had the margins of the grooves closing to form tubes?" In his opinion what was good enough for the cobras ought to have been good enough for the vipers.

Broom was writing prior to establishment of the current neo-Darwinian theory. If alive today, he would ask, "If fitness is measured by number of offspring, and the non-poisonous snakes are more numerous, why did poisonous snakes evolve at all, much less with hypodermic needle-like fangs?"

Snake's egg-breaking Gadget. A South African snake, *Dasypeltis scaber*, has some spines passing from the vertebrae into the gullet. When the snake swallows an egg, these spines cut the shell, enabling the snake to ingest the contents of the egg and eject the shell. "Even if a snake has a million mutations," argued Broom, "it is hard to believe that one of them could be a mutation giving rise to spines from the vertebrae projecting into the gullet and forming teeth—and teeth just where they could be used for cutting egg shells. Nor can we believe for a moment that the snake would die out if it had not those gular teeth. No other snake has them."

Bird coloration. Broom said about that:

> Almost all parrots are showily colored, but they are remarkably well able to hold their own against all their enemies. The little drab colored birds are held to have been thus developed because their inconspicuousness is their safety. But a flock of parrots is as conspicuous as birds could well be. They ought theoretically to have been killed off by hawks, but they get along just as well as the inconspicuous sparrow.

Birds are not color blind. They have better color vision than humans.

The struggle for existence as a guide for evolution was much overrated in Broom's opinion. He noted that it takes place mostly in early life when there is less differentiation, and so he asked, "If a herring lays twenty million eggs in a year and all are eaten by other fish except twenty, are we to assume that those twenty eggs were not eaten because they were fitter or that they would develop into fitter herrings?"

As a paleontologist the Scot was famous for his quick and lucky finds, and as a medical doctor he was familiar with apparently miraculous cures. Among medical doctors this may not be unusual. A survey of eleven hundred doctors discovered that 55 percent had seen cures that appeared to be miraculous. This was reported in 2004 by the Jewish Theological Seminary in New York.

Broom went further than most doctors in regard to spirituality. When he said "spiritual agencies" guided evolution, the plural was not a typographical error. He supposed that *more than one agency* was at work: A good agency designed the songbird, a bad agency the poisonous snake. One of his colleagues referred to Broom's agencies as "good angels and bad angels."

As discussed earlier, Charles Darwin was a disenchanted Christian who rejected the concept of a benevolent God and any other deity. Consequently, he worked on the premise of no Creator. Wallace, reasoning from observable facts, decided there *was* a Creator. Broom, also reasoning from observable facts, decided that there was more than one Creator. We must give Wallace and Broom points for objectivity while bearing in mind that objectivity does not guarantee a correct conclusion.

In 1993, Gould declared the triumph of punctuated equilibrium, but what about the status of neo-Darwinism? Between the Wistar conference of 1966 and Gould's declaration of 1993, there took place a big Macroevolution Conference in 1980, and considering the way it went, this conference would seem to have marked the demise of neo-Darwinism, just as had been sought by the mathematicians. One hundred and sixty biologists assembled in Chicago at the Field Museum of Natural History. The attendants included geneticists, paleontologists, anatomists, and others in the life sciences. Unfortunately, the spirited and often acrimonious proceedings were not published, but there was a piece in *Newsweek,* and *Science* carried a lengthy review by the biochemist Roger Lewin.

The purpose of the conference was to discuss whether big evolutionary changes (macroevolution) result from the accumulation of tiny changes (microevolution) as the neo-Darwinists believed. According to Lewin, the answer for most participants was "a clear No." As *Newsweek* put it: "the story of life is as disjointed as a silent newsreel, in which species succeed one another as abruptly as Balkan prime ministers." Another observer diagnosed the conference as a case of a theory's "spectacular bankruptcy."

The attendants also decided that neo-Darwinism did not explain the long periods of stasis during which no significant evolutionary change takes place. A major proponent of neo-Darwinism, Francisco Ayala, admitted, "We would not have predicted stasis from population genetics, but I am now convinced from what the paleontologists say that small changes do not accumulate." (For details see *Science* volume 80, pp. 883-887.)

"We all went home with our heads spinning," said one participant. Many of the assembled specialists in the life sciences considered the conference an historical event and a turning point in evolutionary biology. But that did not faze the textbook writers.

Considering the manifest defects of neo-Darwinism, the intelligent design people assert that their idea ought be considered in the schools, but the courts always refuse, saying that the intelligent design advocates have a religious motivation. This strikes me as not a valid reason for rejection. Most great scientists, such as Sir Isaac Newton and Louis Pasteur, have believed in a Creator. So did the astronomer Johannes Kepler (1571 to 1630), who also was an astrologer and cast horoscopes. Because he was religious and even an astrologer, shall we reject Kepler's laws of planetary motion? They tell us that most of the time planets revolve around the Sun in elliptical orbits. It surely would handicap our space program if physicists were required to teach the pre-Keplerian idea that planets always move in perfect circles.

Chapter Eight

Non-Darwinian Ideas

And it is not always clear, in fact it's rarely clear, that the descendants were actually better adapted than their predecessors. David M. Raup, Curator of Geology at Chicago's Field Museum of Natural History, 1979.

Who is to say we are not all Martians? Stanford University chemist Richard Zare, press conference, 1996.

In Sir Arthur Conan Doyle's story *The Lost World*, Professor Challenger went to South America twice. On the first trip he found clues to the existence of prehistoric animals surviving on a high, apparently inaccessible plateau, but back in London his report was derided as fraudulent by his colleagues. On the second trip Challenger managed to find a way up to the plateau and, spying some dinosaurs, he exulted (in a movie version) that the scoffers at the Royal Society would have to "eat their hats." To make sure of that, the professor brought back to London a big box. After presenting his new report to a skeptical audience, Challenger opened the box, and out flapped a wickedly grinning, malodorous pterodactyl, which swooped around the room and terrified the assembled scoffers.

Someday many a Darwinian scientist might be eating his hat. In addition to the inconvenient truths we already have noted, the pterodactyl and its fellow pterosaurs have been cited by a distinguished paleontologist as contradictions to Darwinian theory. David Raup did that in an article, "Conflicts Between Darwin and Paleontology," published in the January 1979 *Bulletin* of Chicago's Field Museum of Natural History. In 1979 Raup was Curator of Geology at that museum. Later he became the Sewell L. Avery Distinguished Service Professor at the University of Chicago. Raup must have shocked many a biologist by writing in the *Bulletin*:

> We have no idea why most structures in extinct organisms look the way they do. And, as I already have noted, different species usually appear and disappear from the record without showing the transition that Darwin postulated.

Raup backed up Gould and Eldredge on punctuated equilibrium:

> Instead of finding the gradual unfolding of life, what geologists of Darwin's time, and geologists of the present day actually find is a highly uneven or jerky record; that is, species appear in the sequence very suddenly, show little or no change during their existence in the record, then abruptly go out of the record.

The combative Chicagoan followed that up with a bombshell worthy of Sir Peter Medawar:

And it is not always clear, in fact it's rarely clear, that the descendants were actually better adapted than their predecessors. In other words, biological improvement is hard to find.

That is what Broom said about the transition from reptile to mammal. Mammals may look better than reptiles, at least to our eyes, but did each intervening stage really provide a survival advantage?

Another of Raup's observations helps to explain why invalid examples of evolution—what Gould called "weak and foolish speculation"—continue for so long in the textbooks. By 1979 a quarter million fossils had been collected, but the record remained gappy, and, "ironically," said Raup, "we have even fewer examples of evolutionary transition than we had in Darwin's time." As one lost example Raup mentioned the horse, whose evolution turned out to be "much more complex and much less gradualistic." The Darwinists were better off when they had fewer fossils. Then their leader could complain in *The Origin of Species,* "What a paltry display we behold!"

As a puzzling example of transition, Raup discussed the pterosaurs. There were many sizes of these flying animals, ranging from that of a sparrow to the size of a small airplane. The largest of them, the pteranodon, had a wingspan of about fifty feet and, appropriately, was discovered in Texas. The first pterosaur remains were found in 1784, and more than three thousand specimens of eighty-five species have turned up. Yet no fossil can be identified as the pterosaurs' ancestor. In 1996 a symposium on pterosaurs decided that the animal's biggest mystery was its unknown origin, according to the *New York Times* of October 22, 1996.

To Raup the pterosaurs' disappearance was just as puzzling as their origin. They were successful predators for 135 million years. To explain their extinction, Darwinism tells us that there must have been something wrong with the pterosaurs, but according to Raup, we have no evidence of that. As for natural selection in general, it was Raup's opinion that natural selection does take place, but for lack of evidence in the fossil record, one cannot tell how important it has been—"whether it was responsible for 90 percent of the change we see, or 9 percent, or .9 percent."

In 1979, Raup's skepticism about the importance of natural selection as a cause of extinction was just a little ahead of its time. In the very next year came big news. It was reported that the pterosaurs vanished along with the dinosaurs, about sixty-five million years ago, as the result of a mass extinction caused by the impact of a large extra-terrestrial object or objects. The collision threw an estimated *12,000 cubic miles* of dust and debris into the sky, and in theory the resulting darkness killed much of the planet's plant life and then, for the lack of herbaceous food, most of the animals. That was the end of the Cretaceous period. Overall it has been estimated that 80 percent of species died off. Some small mammals were among the survivors.

Evidence for an impact extinction was not known when Raup published his article, but hard data in support of that were published in the journal *Science* in 1980 and this shocked a disbelieving scientific community.

Consider the background. Before the ugly fact of catastrophe came along, the Darwinists were able to float any sort of story about extinction that they happened to like. At Wistar, Ernst Mayr stated flatly that the "dinosaurs, the pterodactyls, and so many other types" became "so specialized in certain ways" that they got "themselves into a dead end and eventually became extinct." The "dead end" theory was new to me. Previously, I had heard that the dinosaurs were stupid and never learned to cooperate with each other. Concerning cooperation, however, fossil discoveries have suggested that some dinosaurs roamed in herds and some hunted in packs. As for other theories of the dinosaurs' extinction, Raup recounted climatic cooling, constipation, sterility, and unspecified "wilder ideas."

One college level textbook asserted that a change in climate caused the demise of the dinosaurs, and this book specifically rejected meteorite impact as a possible cause. Impact suggestions for mass extinction had been made for some time. The irrepressible Otto Schindewolf had proposed a supernova explosion as responsible for the Permian mass extinction of 250 million years ago. That was the biggest extinction in our planet's history, with perhaps 96 percent of species destroyed (what a close call). In 1970, the Canadian paleontologist Digby McClaren suggested a meteorite impact as a possible cause of the Devonian extinction about 365 million years ago. Harold Urey, who had tried to create life by electrifying a mixture of gases, in 1973 speculated in *Nature* that the impact of a comet had caused the Cretaceous extinctions.

But none of those theorists had evidence. It simply was their opinion that some really huge cataclysm had occurred, and no possible terrestrial cause seemed sufficient. Lacking physical evidence, the extraterrestrial ideas received little notice, even though Urey was a Nobel Prize winner (in chemistry). The above mentioned textbook declared that "sensible scientists" ought to look for ordinary causes for events rather than extraordinary causes that would be difficult to prove. Deeply ingrained was Sir Charles Lyell's theory of gradual change, with the present serving as the key to the past.

"Sensible scientists" gave us many unsubstantiated, yet better received, explanations for extinctions. One might be reminded of the old joke: "If you lost it over there, why are you looking for it over here?" Answer: "Because the light is better over here." Darwinists prefer explanations that can be seen in the light of their theory.

As another black mark against mass extinction from space, catastrophism was the Bible-friendly explanation of extinctions that scientists favored prior to *The Origin of Species*. What a bitter pill for the Darwinists was the return of catastrophism. To think that the "dukelet" of Argyll could have been nearer the truth than Lyell, Huxley, and Darwin!

Even with evidence, the impact theory for long received no respect. When the Alvarez team of scientists reported that the dinosaurs were part of a mass extinction caused by the impact of an extraterrestrial object or objects, this, according to Raup, was so incredible as to be "like suggesting that the dinosaurs had been shot by little green men from a spaceship."

As another impediment to the theory's acceptance, nobody on the team was a biologist. Luis Alvarez was a physicist, his son Walter a geologist, Helen Michel a chemist, and Frank Asaro a chemist. Like Louis Pasteur, whose work was rejected by medical doctors, the Alvarez team did not have the credentials needed for upsetting a whole range of dogma in somebody else's field of competence. A dinosaur expert at the University of Colorado gave the catastrophists a piece of his mind

> The arrogance of these people is simply unbelievable. They know next to nothing about how real animals evolve, live, and become extinct. But despite their ignorance, the geochemists feel that all you have to do is crank up some fancy machine and you've revolutionized science. The real reasons for the dinosaur extinctions have to do with temperature and sea level changes, the spread of diseases by migration, and other complex events. But the catastrophe people don't seem to think such things matter. In effect, they're saying this: "We high-tech people have all the answers, and you paleontologists are just primitive rock hounds."

Wounded vanity? Raup defended the "high tech people" as respectful toward the rock hounds and eager to learn.

The Alvarez discovery was serendipitous, like Columbus encountering America while looking for Asia. When studying paleomagnetism in Italy, Walter Alvarez examined a thin layer of exposed clay that marks the boundary between the plant and animal life of the Cretaceous period from the rather different *flora* and *fauna* of the Tertiary period, and thinking the clay might offer a clue to the extinction of Cretaceous organisms, the geologist had it analyzed. Chemist Frank Asaro then discovered in the material an unusually large amount of the element iridium, which is rare in the earth's crust but commonly found in meteorites. Although geologists were taught that the concept of catastrophic change was unscientific, the iridium discovery led to the hypothesis of an asteroid or comet impact's having been responsible for the mass extinctions that ended the Cretaceous. The one-centimeter-thick layer of clay was the fallout, and it blanketed the world.

Eventually, among other evidence, what appeared to confirm the impact explanation was the discovery of a 112-mile wide crater, 65 million years old, at the edge of the Yucatan Peninsula. (There also has been identified a somewhat older 25-mile wide crater in Iowa.) In 1995, there was reported the "smoking gun": a fragment of the actual doomsday rock. Geophysicist Frank Kyte discovered a tiny iridium-rich chip in a core sample from the Yucatan crater.

Notwithstanding the accumulation of evidence for extinction by impact, some geologists argued that the Cretaceous extinction was caused by the eruption of volcanoes.

In the history of our planet there have been five major mass extinctions, the Big Five, and unknown number of smaller ones. David Raup, along with his partner John J. Sepkoski, produced statistical evidence that the Big Five occurred at 26-million-year intervals, and the two paleontologists hypothesized that our planet periodically has been struck by large objects from outer space. Adding some credibility to that idea is the fact the surface of the Moon is covered with impact craters. Craters on the Earth usually are eroded into the landscape by weather. The airless Moon, of course, has no weather.

As for why our planet gets pelted on a regular basis, it has been suggested that the periodic approach of an unknown planet or star disturbs comet orbits in the distant Oort cloud of comets or comets in a closer assemblage, the Kuiper Belt. Some of the comets then would be pushed in all directions including ours. The term *Nemesis* has been given to the unknown star or planet (also dubbed Planet X). As yet no Nemesis body has been identified. Slightly more than half of the Sunlike stars in our galaxy have companion stars, according to the Smithsonian-Harvard Center for Astrophysics, and so it would not be unusual for our local star to be one of a pair orbiting around a common center of gravity.

Such a science fiction scenario was too much for sober scientists. In 1985, the *New York Times* reflected the general reaction of the scientific community in an editorial that said, "Astronomers should leave to astrologers the task of seeking the cause of earthly events in the stars." Subsequently *Nature* presented a skeptical paper and editorially agreed with the contents. *Science,* the leading American journal, judged it was too early to decide on either the periodic extinction or the Nemesis theory.

There is another kind of extraterrestrial influence, and luckily this is a good one. A giant planet, Jupiter, stands guard. In 1994 its massive gravity pulled in the big Shoemaker-Levy comet, which crashed spectacularly. But inevitably there are objects that do not get short-stopped, as in the case of the Tunguska Event. In the summer of 1908, a large object exploded in the sky over Siberia and flattened 500,000 acres of pine forest.

Now that we have become aware of dangers from outer space, the U.S. Congress has ordered the National Aeronautics and Space Administration (NASA) to search for "near Earth objects" (NEOs) that might strike our planet. The plan is for NASA to find by the year 2020 at least 90 percent of NEOs more than 460 feet in diameter. Unfortunately, objects on a collision course are the most difficult to detect since they present no relative motion against the stars. NASA hopes to improve its observational ability by building a special observatory on Earth or in a Venus orbit, but at this writing funds are not available.

What could be done about an approaching doomsday rock is being debated. There is a nuclear option (setting off a bomb) and a tractor option (pushing the interloper with spacecraft or by attaching rockets). A serious complication is that Shoemaker-Levy approached Jupiter as a swarm of more than twenty objects. The comet evidently had broken up on an earlier, 1992 approach to the big planet, although—and here is a cautionary tale—its existence was not noticed until 1993. The comet was discovered by Eugene Shoemaker, his wife

Carolyn Shoemaker, and David Levy. (Carolyn Shoemaker holds the record for comet discoveries—thirty two. She also has spotted more than eight-hundred asteroids.)

If prior to learning about mass extinctions caused by objects from outer space, one could not be sure of whether evolutionary change resulted from natural selection 90 percent of the time, 9 percent of the time, or .9 percent of the time, what is the likelihood now? It must be substantially less. How much less? Raup believes that *most* extinctions have resulted, as in the case of the dinosaurs, not from bad genes but from bad luck, that is, causes having nothing to do with terrestrial adaptation. This opinion, he says, "is shared by many of my colleagues even though a majority of paleontologists and biologists still subscribe to the Darwinian view of evolution, that of a constructive force favoring the most fit." Raup himself remains convinced that natural selection is the "only viable, naturalistic explanation we have for sophisticated adaptations like eyes and wings."

A more skeptical view is taken by Kenneth J. Hsu (pronounced like the French word *jus*). Hsu is the celebrated Chinese geologist who by drilling the floor of the Mediterranean Sea discovered that the sea dried up 5 million years ago. Hsu's geochemical research also has provided evidence for a global catastrophe at the time of the dinosaur extinction.

In the Chinese geologist's opinion, extinction by natural selection is a matter of faith rather than science. As he said in a letter to me, Hsu once challenged a conference of biologists "to give one single piece of fossil evidence that any species other than *Homo sapiens* had exterminated or directly caused the extinction of another." Hsu's challenge aroused such fury that the conference turned into a "public inquisition." But according to Hsu, nobody provided the requested evidence. "Few of them seemed to have read Darwin's *Origin of Species*," said the geologist (making a charge that the Darwinists routinely throw at the creationists). "They took everything on faith." (There's that word again.)

To be fair to the orthodox biologists, I can mention that grey squirrels have been replacing the red squirrels in England. However, the grey squirrels did not evolve in England. They were introduced by humankind from North America, and they are spreading a lethal virus.

In support of Hsu the July 20, 2007, issue of *Science* reported startling news about the emergence of dinosaurs. Instead of the dinosaurs out-competing and thereby eliminating their precursors as they evolved in what is now New Mexico, their precursor species survived for another 15 to 20 million years. According to Randall Irmis, a co-author of the report in *Science*, "*If there was any competition* [emphasis added] between the precursors and dinosaurs, then it was a very prolonged competition."

Charles Darwin had a very dark vision of life—"Nature, red in tooth and claw" in the words of Tennyson. Huxley called animal life a "gladiators' show." That view of nature became dominant, but according to Prince Kropotkin it was very incomplete. During the years of the geologist's traveling in Siberia, he observed that most animals lived a very sociable kind of life and that they helped each other in many ways. Wild horses, for example, formed herds and when a predator approached, several studs would get together and drive it off. Only horses detached from the herd fell prey. Between species there was conflict, said the prince, but within species he saw a great deal of sociability and mutual aid. (Yet we must note that often competition does take place between groups and individuals within the same species.)

From England, Kropotkin reported cooperation, perhaps even altruism, from a surprising source. At the Brighton aquarium a crab fell on its back and could not right itself because an iron bar happened to be in the way. Two crabs came to help, but even they were defeated by the iron bar. One crab then went away and came back with two more crabs. Joint efforts continued, with no success, for more than two hours at which time Kropotkin left. In his book *Mutual Aid* Kropotkin asserted, "Sociability is as much a law of nature as mutual struggle," and he raised this question: "Who are the fittest, those who are continually at war with each other, or those who support one another?"

Cooperation among animals sometimes blends into altruism, and we can find many reports of that:

Dogs adopting orphaned animals of other species including cats, squirrels, ducks, and even tigers. (As I write this chapter a terrier in Australia is being hailed a hero for guarding four kittens during a house fire.)

Dolphins supporting sick or injured animals for hours at a time.

Male baboons covering the rear as the group retreats from a predator.

Chimpanzees offering food when shown a certain gesture.

As Marx used natural selection theory to justify violent revolution, Kropotkin used his mutual aid theory to propose that people could live best in communal societies and with no need for an oppressive central government.

Although a longtime revolutionary, the "anarchist prince" was deeply disappointed to see the Bolsheviks take over the governance of Russia with their tyrannical methods and amoral philosophy. Bolshevik amorality could be mind-boggling even in relatively mundane matters. After Lenin died his widow tried to offer some advice to the ruling Politburo, and a Stalin henchman, Lazar Kaganovich, told her to desist or she would be replaced with another widow. Kaganovich, a man of infinite evil, played a major role in the Ukrainian holocaust and other mass murders. He died in 1991 at the age of 97.

As mentioned earlier, the cell biologist Lynn Margulis subscribes to a symbiogenetic evolution, which emphasizes cooperation rather than Darwin's dog-eat-dog competition. It is claimed to provide a more credible source of genetic variation than the neo-Darwinian notion of random genetic mutation.

Symbiogenesis was proposed first by a Russian botanist, K. S. Merechovsky, who perhaps drew some inspiration from Kropotkin. In

1926, the botanist published *Symbiogenesis and the Origin of Species*, and it was he who first came up with the idea that the nucleated cell resulted from one bacterium invading another. In 1927, the American anatomist Ivan Wallin, a son of Swedish immigrants, suggested that the action of bacteria might be the fundamental cause of evolution. His book was titled, *Symbionticism and the Origin of Species.*

Margulis, too, urges the importance of bacteria in evolution, and she says that bacteria have provided not just the nucleus but most of the DNA found in eukaryotic cells. The combining of bacteria is called "endosymbiosis," and Eldredge has hailed it as "probably the grandest idea in modern biology."

In 2000, the journal *Trends in Microbiology* reported that genome mapping had found evidence of bacteria carrying genetic material from one evolutionary family tree to another even at high levels. The paper "Horizontal gene transfer and the origin of species: lessons from bacteria" was contributed by F. de la Cruz and J. Davies. Another striking example of gene transfer was reported in 2008. According to scientists at the University of Ontario, three species of fish share an identical protein needed to prevent their blood from freezing in cold water, even though these species do not share a recent common ancestor. Fish sperm is transmitted externally, and it is surmised that sometimes a fertilized egg receives a bit of the wrong DNA. (This is not hybridization; note that a fertilized egg is specified.)

Now, how about an explanation for why mammals and the invertebrate squids have similar eyes? We humans and the squids are less related than those three species of fish.

Raup retains some confidence in natural selection, saying it is needed in order to account for complex objects such as wings and eyes. Yet the Darwinists have no good explanation for either the feathered wings of birds or the non-feathered wings of insects (remember Mayr's list in Chapter Six?). As for the human eye, Darwin said of it in *The Origin of Species*:

> To suppose that the eye with all its inimitable contrivances for adjusting the focus to different distances, for admitting different amounts of light, and for the correction of spherical and chromatic aberration, could have been formed by natural selection, seems, I freely confess, absurd in the highest degree.

Nevertheless, Darwin argued at length for that as having taken place. To capsulize the argument: Admittedly, he said, one can find no linear progression for evolution of the eye, but looking around at the *various* animals, from lowest to highest, one can see an overall *trend* from optically sensitive nerve to fully formed eye. In the following year, 1860, Darwin wrote to Asa Gray: "To this day the eye makes me shudder."

Birds were very troublesome, too. Darwin confessed to Gray: "The sight of a feather in a peacock's tail, whenever I gaze at it, makes me sick!" It would seem that any evidence of a divine plan, as in the case of the eye or the peacock, gave the scientist a shock and might have contributed to his poor health.

Obviously, the peacock with his six-foot train, which fans out to display a multi-colored, iridescent constellation of eyes, did not come about as a means of hiding from predators or as a means of rapid escape. The size and showy appearance of the bird are survival handicaps. So what did Darwin come up with as the mechanism for the peacock's design? Sexual selection. The peahen, he decided, prefers to mate with good-looking males (as do other females), and so over a period of time, *voila*, the peacock, an amazingly ornate bird surmounted by a fancy crest.

The sexual selection explanation appears to be accepted unquestioningly by scientists. Whenever I ask one about the peacock, I'm told something like, "The female peafowl are very finicky when it comes to mating." That leaves me just as unsatisfied as when I'm glibly told that feathers evolved from scales. *How could that have happened?* I want to know. Concerning the Peahen Solution I respond, Okay, the females of the peafowl are extraordinarily finicky. So in medieval Italy was the Medici family. But members of the Medici family were not great painters or sculptors, and to satisfy their esthetic cravings they had to hire Michelangelo. Yet finicky peahens, gaggles of them, over millions of generations designed the peacock? Except to save Darwinism, why would anybody believe that? Where is the evidence?

Dun-colored birds, such as the wren, manage to mate. Why are they not so choosy? As a logical Darwinian answer, extraordinary finickiness about mating would be a survival handicap.

The design of the peacock came from somewhere. Was it produced by generations of artistic peahens? Or, as Mivart would insist, by the power of a Mind? Like Mivart, most proponents of intelligent design have been religious. An early example was the author of the Bible's Psalm 19 (attributed to David), which tells us, "The heavens declare the glory of God, and the firmament showeth His handiwork." Creationists might well view the peacock as the Creator's joke on secular evolutionism.

As another intelligent design argument, the biochemist Michael Behe in his book *Darwin's Black Box* has described living organs that he says are irreducibly complex. In other words, they, like a mousetrap, *cannot function at all until fully assembled.* One cannot catch a mouse with just the platform, just the spring, or just the hammer. Everything has to be together for the contraption to work. For that reason, says Behe, an irreducibly complex living organ could not have evolved piece by piece through natural selection. We are reminded of Mivart's wing and Murray Eden's hemoglobin. As one example of irreducible complexity, Behe in a *Wall Street Journal* article of October 29, 1996, cited the bacterial flagellum:

> a rotary propeller, powered by a flow of acid, that bacteria use to swim. The flagellum requires a number of parts before it works—a rotor, stator, and motor. Furthermore,

genetic studies have shown that about 40 different kinds
of protein are needed to produce a working flagellum.

In rebuttal, Darwinists say that the bubonic plague bacterium has a similar looking apparatus featuring a needle that is used to inject poison into the bacterium's host, and so the more complex flagellum motor could have evolved from that. The intelligent design people have a response. They say that the bubonic plague bacterium has a full set of genes for making the flagellum motor, and the sequence of genes indicates that the flagellum part came first.

Let us move on to a clearly secular sort of intelligent design that was proposed by Francis Crick, the molecular biologist who got frustrated by the origin of life problem. Crick was not religious at all. He described himself as an agnostic with "a strong inclination toward atheism." The DNA Nobelist accepted a fellowship at Cambridge's Churchill College with the explicit understanding that the college would not build a chapel, and when a large donor persuaded the college to change its mind, Crick resigned. His last years were spent in research at the Salk Institute in California. The Britisher felt "at home" in Southern California. Upon his death, his body was cremated and the ashes scattered in the Pacific Ocean.

In his book *Life Itself* Crick raised a lot of eyebrows by suggesting that life might have been dispatched here in a space ship by an advanced civilization on another planet. Frozen microorganisms, he said, could survive a trip of ten thousand years. Of course, that does not explain how the very first life came about. But it could explain how life got underway on *this* planet and perhaps explain also the Cambrian Explosion.

A different theory about life's coming from outer space was published in the same year as Crick's book, 1981, by the astronomer Sir Fred Hoyle. In his book *Evolution from Space* Hoyle took up the idea that life permeates outer space. This was not a new idea. It had been around for more than a hundred years. Life from space was proposed in the nineteenth century by the German physicist H. E. Richter, and then by the British physicist William Thompson, Lord Kelvin, the man who devised the Kelvin scale of absolute temperature. Thompson thought that many living worlds had preceded ours and that their collisions would have produced life-bearing meteorites.

Another means of transmitting life was proposed by the Swedish chemist, Svante Arrhenius, who won a Nobel Prize for explaining how the electric storage battery works. Arrhenius believed that spontaneous generation was impossible, and so in 1907 he published a book, *Worlds in the Making*, in which he proposed that bacteria of a certain size could be propelled through space by waves of starlight. Could they survive the extreme cold of outer space? Evidently some could. Experiments in London had shown that certain bacterial spores retained their germinating power after exposure to the temperature of liquid hydrogen. Critics said the bacteria would be destroyed in outer space by ultra-violet light. In support of Arrhenius, Hoyle reported that outer space contains large quantities of graphite dust, and he argued that spores attached to such particles would be protected from ultra-violet rays.

Hoyle was an intelligent design man, and as a significant property of bacteria, he cited the ability of the "radioresistant micrococci" to survive doses of X radiation "many millions of times greater than they could ever have received in the natural environment." The bacteria are not undamaged. They are heavily damaged, but they are able to repair the damage with "amazing efficiency." Hoyle then made this point concerning the bacteria's ability to survive such powerful radiation

> The question was, and still is, how the highly specific enzymes needed for the repair process came to be present in the bacteria, since in the terrestrial environment there are no X-rays that could exert selective pressure for their development.

In Hoyle's view, these very strange bacteria could not have evolved by some hit-or-miss process on this planet.

The astronomer favored a vehicle different from dust for the space-traveling microorganisms, and that was the comet. Some scientists have speculated that our planet, when young, received all of its water from comets. Hoyle estimated that a thousand tons of cometary material still drift into the earth's atmosphere every year. He believed that comets carry life through space, seeding planets with it, and that these icy bodies not only brought the first life to our planet but also have delivered—and still deliver—bacteria and viruses that cause disease.

As one possible example of disease, Hoyle cited the Spanish influenza pandemic of 1918 to 1920. This influenza virus killed an estimated 50 to 100 million people, many of them young healthy adults. Hoyle contended that the disease spread around the world so fast, in those days before commercial air travel, that viruses must have rained down from the sky.

Hoyle died in 2001, but a collaborator of his, Chandra Wickramasinghe, continues research on their theory. In 2003, Wickramasinghe and some other British scientists suggested that the deadly severe acute respiratory syndrome (SARS) virus, described as "unexpectedly novel," was inflicted upon us by a comet.

The life from space theories attributed to Hoyle and Arrhenius both are called "panspermia." Crick's theory is called "directed panspermia."

Hoyle believed that intelligent design had produced both the universe and the life within it. To Hoyle the insects were deliberately designed as colonizers able to survive almost any sort of hardship, including nuclear radiation. They appeared abruptly on our planet 400 million years ago, and no mass extinction has wiped them out.

For many years, Hoyle was disappointed that his ideas about life coming from space were either ignored by the scientific community or derided as "downright silly" and "Hoyle's howler." Like the duke of Argyll, the astronomer accused the Darwinian biologists of spreading "dogma" in a "campaign of propaganda." Hoyle and his partner

Wickramasinghe made that accusation in their 1993 book *Our Place in the Cosmos*.

The situation changed drastically just a few years later. In 1996, scientists connected to NASA became very excited about a meteorite, found in the Antarctic, that looked as if it had harbored life in the form of microorganisms. Before the initial news conference, a source "close to the agency" told CNN, "I think it's arguably the biggest discovery in the history of science." Carl Sagan called the discovery "glorious." It also represented an opportunity to seek research funding.

The potato-shaped rock was one of twelve believed to have come from Mars, perhaps having been knocked loose from the planet by an asteroid. Chief evidence for the meteorites' Martian origin were traces of gas that were chemically the same as the Red Planet's distinctive atmosphere, which had been analyzed in 1976 by NASA's Viking Lander spacecraft. The rock supposedly bearing evidence of life arrived an estimated thirteen thousand years ago, probably after wandering for millions of years in space. A NASA research team felt confident that certain marks on the rock indicated the former activity of living organisms. On the assumption that life first came to this planet from Mars, Stanford University chemist Richard Zare raised the rhetorical question, "Who is to say we are not all Martians?" Several years later, the Mars rock issue was debated at a conference in Houston, and one expert on meteorites summed up the pros and cons by saying, "It's a definite maybe."

Hoyle's belief that the universe is the product of intelligent design in part was based on facts similar to those advanced some years later, in 1988, by the astrophysicist Stephen Hawking. In his book *A Brief History of Time*. Hawking observed that the universe contains many fundamental numbers, such as the size of the electric charge of the electron, and said "The remarkable fact is that the values of these numbers seem to have been very finely adjusted to make possible the development of life." He explained

> For example if the electric charge of the electron had only been slightly different, stars either would have been unable to burn hydrogen and helium, or else they would not have exploded [releasing the heavier elements, such as carbon and oxygen, that are needed for life as we know it] . . . Most sets of values would give rise to universes that, although they might be very beautiful, would contain no one able to wonder at that beauty.

On the other hand, conceivably the fundamental numbers producing life came about by accident in one of many randomly emerging universes. Hawking remarked on how man always has put himself at the center of things, but keeps learning otherwise. Man has learned that the Earth is not the center of the universe, neither is the Sun, and neither is our galaxy. The Earth, said the astrophysicist, "is just a medium-sized planet orbiting around an average star in the outer suburbs of an ordinary spiral galaxy." Hawking found it hard to believe that the "whole vast construction exists simply for our sake."

Carl Sagan used to make that same point on television, but after he died astronomers learned that it would have been no great honor for our planet to have been placed in the center of the galaxy. Located there is a black hole that sucks in everything around it. Here in the suburbs we are far from that danger, we are much less crowded by other stars and planets, and we even have Jupiter as a guardian.

Then, as icing on the cake, our solar system occupies an area of space relatively empty of galactic dust, which probably was cleared out long ago by the explosion of a supernova. This "bubble" or "space chimney" provides a clear view far into the depths of the universe, a very convenient arrangement for cosmologists like Hawking and Sagan, and for the progress of human knowledge. The clear view also ought to help us find incoming mail.

Perhaps someday we will be able to decide scientifically whether intelligent design has played a role in the formation of our universe and the origin and development of life. Until then, some people will believe in it and some will not.

Evidently, even for the most committed secular evolutionist it is difficult to pull away entirely from the religious sensibility. Thomas Huxley liked to sing psalms of a Sunday, and as a member of the London School Board, Huxley applied his eloquence to supporting Bible study in the schools in order to teach moral behavior.

To Charles Darwin the Arthur S. Sullivan hymn "Will He Come?" was a "never failing enjoyment," according to son Francis. Darwin valued the local Reverend Innes as one of his best friends, and the two worked together on parish matters, even though the scientist did not attend church. Darwin said of the rector, "[He] and I have been fast friends for thirty years. We never thoroughly agreed on any subject but once and then we looked hard at each other and thought one of us must be very ill." In 1881, nearing death, Darwin often asked for the works of Bach and Handel to be played on the piano.

Coincidentally, a thirty year period comes up in connection with Stephen Gould. Although a Jewish agnostic, Gould for thirty years sang Handel's *Messiah* and other religious works as a member of the Boston Cecilia chorus.

One of the nation's best biology textbook writers, the late Neil A. Campbell, for the 1993 edition of his *Biology: Concepts and Connections*, interviewed Mayr and Gould, asking each the hypothetical question: What would you would want to know if you had the opportunity to talk to Charles Darwin?

"You will be very much surprised when you hear my answer," said Mayr. "I would ask him about his relation to religion." The professor, then retired, wanted to know "what went on in [Darwin's] mind" about that. He knew that Darwin lost his faith in a personal God, and he believed that this took place prior to Darwin's concept of natural selection. "So," said Mayr, the claim that biology and a belief in natural selection is dangerous because it may make you lose your faith in God is not substantiated." He added, "That's a very important thing to know."

But Darwin's personal example was not the whole story, and Thomas Huxley filled a growing, Darwin-inspired need when he invented the word the word *agnostic*. Did Mayr's rather desperate argument betray a sense of guilt for the nature of his long career? Did he mean to encourage a belief in God?

Asked what he would talk to Darwin about, Gould, too, said religion. He wanted to know about Darwin's "personal religious feelings, which he was very cryptic about." Gould was a great admirer of Charles Darwin and frequently wrote about him. One would think the professor must have known about his hero's agnosticism. In the autobiography Darwin established clearly his true position on religion saying, "I cannot pretend to throw the least light on such abstruse problems. The mystery of the beginning of all things is insoluble by us, and I for one must be content to remain an Agnostic."

Either Gould missed that or, unsatisfied, he just wanted to press Darwin for more—as did Mayr.

Concerning the existence of a Creator, Stephen Hawking in *A Brief History of Time* had this to say: "The whole history of science has been the gradual realization that events do not happen in an arbitrary manner, but that they reflect a certain underlying order, which may or may not be divinely inspired."

Physicists can use words like *divinely*, but not the biologists. Their profession has a prejudice.

As for what to teach in the schools, why not tell the truth? We could start by quoting Mayr. In his book *One Long Argument* (page 143) he said, "Darwinism is not a simple theory that is either true or false but is rather a highly complex research program that is being continually modified and improved." Considering that statement and the many evolutionary problems and mysteries that we have discussed, it is obvious that our scientists do not have a complete theory of evolution. What they have is a partially successful attempt to construct one (owing much to Wallace's Sarawak Law). Clearly lacking is an adequate source of design. The true situation ought to be acknowledged in the textbooks, and speculative examples of Darwinian evolution—or any other kind of evolution—ought to be labeled as such. If the example is Lamarckian, let us say it is Lamarckian.

Gradual evolution by natural selection has been rejected repeatedly in the scientific literature, but Darwinism, like Dracula, keeps coming back. The science establishment likes it, textbooks present false evidence for it, and a series of famous biologists have appeared to support the theory even though they did not believe in it. The fear of creationism has warped generations of thinking, and telling the truth has been frowned upon if not excoriated. Science thrives upon accurate information, original thinking, and free discussion. Supporting those principles ought to be praiseworthy, not a firing offense.

Index

Wald, George, 35, 204-6
Wallace, Alfred Russel, 62-70, 72-73,
    76-81, 83, 85, 96, 100, 136,
    146, 159, 166, 173, 221, 251
  Moral Progress, 76
  "On the Law Which Has
    Regulated the Introduction
    of New Species", 62
  "On the Tendency of Varieties to
    Depart Indefinitely from the
    Original Type", 72
Wallace Line, 77
Wallin, Ivan, 238
  *Symbionticism and the Origin of
    the Species*, 238
Walsh, John E., 140
  Unraveling Piltdown, 140
War of the Worlds, The (Wells), 114
Wells, H. G., 114
  Time Machine, The, 114
  War of the Worlds, The, 114
White, George, 113
Wickramasinghe, Chandra, 245
Wilford, John Noble, 157
"Will He Come?" (Sullivan), 249
Wilson, Woodrow, 105
wind friction, 185
"Winning by a Neck", 174

wire-pullers, 43, 77, 93
Wistar Institute, 201
Wizard of Sussex. *See* Dawson,
    Charles
Woodward, Arthur Smith, 139,
    141-48, 150-51, 155
  Earliest Englishman, The, 139
Worlds in the Making (Arrhenius),
    243
Wright, Robin, 215
Wright, Sewall, 198, 201, 204-5

X

X Club, 43, 51, 87, 89–91, 93, 95, 97,
    99, 101, 136, 160, 187, 190
X Clubbers, 86, 91, 94

Y

Yaroslavsky, Emelyan, 169
York, Alvin, 117, 122
Yushchenko, Viktor, 171

Z

Zare, Richard, 224, 246
Zoonomia (Erasmus Darwin), 49

LaVergne, TN USA
12 May 2010
182522LV00001B/103/P